W9-AGW-592

Springer Series in Computational Mathematics

6

François Robert

Discrete Iterations

A Metric Study

With 126 Figures

Translated by Jon Rokne

Springer-Verlag
Berlin Heidelberg New York Tokyo

Professor François Robert
University of Grenoble, Institut IMAG
B.P. 68
F-38402 Saint Martin d'Heres, France

Translated from the French manuscript by
Professor Jon Rokne
Department of Computer Sience
University of Calgary
Calgary, Alberta, Canada, T2N-IN4

Mathematics Subject Classification (1980):
15A, 65F, 65H, 68D, 68E

ISBN 3-540-13623-1
Springer-Verlag Berlin Heidelberg New York Tokyo
ISBN 0-387-13623-1
Springer-Verlag New York Heidelberg Berlin Tokyo

Library of Congress Cataloging-in-Publication Data
Robert, François, 1939–
Discrete iterations.
Bibliography: p.
Includes index.
1. Iterative methods (Mathematics) I. Title. QA297.8.R63 1986 519.4 85-27949
ISBN 0-387-13623-1 (U.S.)

Printed in Germany

Typsetting, printing and bookbinding: Universitätsdruckerei H. Stürtz AG, Würzburg
2141/3140-543210

Ce livre est dédié à la mémoire de Noël Gastinel (1925–1984)

"If the human brain was so simple that we could understand it, then *we* would be so simple that we could not" (L. Watson)

Acknowledgements

This book is dedicated to the memory of Professor Noël Gastinel (1925–1984) in recognition of his seminal contribution in the domain of Computational Mathematics, and of his leading role in furthering the scientific work of many people. He has been a master for many of us, and I personally remember him with great affection and gratitude.

I would like to direct my thanks to my colleagues who encouraged me to publish this monograph, and who helped me with their remarks: in particular, Professor R.S. Varga, and the members of our working group in Grenoble, especially Michel Cosnard, Eric Goles, Yves Robert, and Maurice Tchuente. For me, the writing of this book is a part of our common history. I am also indebted to Professor Philippe Ciarlet and to Françoise Fogelman-Soulie for their kind help.

The CNRS, IMAG/TIM3 and INPG/ENSIMAG are to be thanked for the excellent working conditions within the "Equipe d'Algorithmique Mathématique" in Grenoble.

The interactive process of translating the book became the occasion of a very confident and enriching collaboration with Professor Jon Rokne. I would like to thank him very much for his work and for the care he took, as well as for his help and suggestions. He also let me benefit from his experience. Because of the time we spent on this book, it has now become ours jointly. I am also very grateful to Mrs Patricia Dalgetty, of the University of Calgary, for typing the manuscript with much skill and care. I am indebted to the staff of Springer-Verlag for their fine work throughout the publication process. Last but not least, I would like to thank my wife Irène, who carefully checked the proofs.

Grenoble, April 1986 F. Robert

Table of Contents

Introduction I

1. Discrete Iterations and Automata Networks: Basic Concepts . . . 1
   1. Discrete Iterations and Their Graphs 1
   2. Examples 3
   3. Connectivity Graphs and Incidence Matrices 7
   4. Interpretations in Terms of Automata Networks 9
   5. Serial Operation and the Gauss-Seidel Operator 11
   6. Serial-Parallel Modes of Operation and the Associated Operators 17

2. A Metric Tool 27
   1. The Boolean Vector Distance d 28
   2. Some Basic Inequalities 29
   3. First Applications 31
   4. Serial-Parallel Operators. An Outline 36
   5. Other Possible Metric Tools 40

3. The Boolean Perron-Frobenius and Stein-Rosenberg Theorems . . 43
   1. Eigenelements of a Boolean Matrix 43
   2. The Boolean Perron-Frobenius Theorem 51
   3. The Boolean Stein-Rosenberg Theorems 53
   4. Conclusion 55

4. Boolean Contraction and Applications 57
   1. Boolean Contraction 58
   2. A Fixed Point Theorem 58

3. Examples 62
4. Serial Mode: Gauss-Seidel Iteration for a Contracting Operator 65
5. Examples 66
6. Comparison of Operating Modes for a Contracting Operator . . 69
7. Examples 70
8. Rounding-Off: Successive Gauss-Seidelisations 76
9. Conclusions 78

5. Comparison of Operating Modes 79
1. Comparison of Serial and Parallel Operating Modes 80
2. Examples 88
3. Extension to the Comparison of Two Serial-Parallel Modes of Operation 90
4. Examples 91
5. Conclusions 92

6. The Discrete Derivative and Local Convergence 95
1. The Discrete Derivative 95
2. The Discrete Derivative and the Vector Distance 99
3. Application: Characterization of the Local Convergence in the Immediate Neighbourhood of a Fixed Point 103
4. Interpretation in Terms of Automata Networks 107
5. Application: Local Convergence in a Massive Neighbourhood of a Fixed Point 108
6. Gauss-Seidel 115
7. The Derivative of a Function Composition 117
8. The Study of Cycles: Attractive Cycles 121
9. Conclusions 129

7. A Discrete Newton Method 131
1. Context 132
2. Two Simple Examples 135
3. Interpretation in Terms of Automata 137
4. The Study of Convergence: The Case of the Simplified Newton Method 138
5. The Study of Convergence, The General Case 145
6. The Efficiency of an Iterative Method on a Finite Set 153
7. Numerical Experiments 154
8. Conclusions 165

General Conclusion 166
Appendix 1. The Number of Maps of $\{0,1\}^n$ into $\{0,1\}^n$ 167
Appendix 2. The Number of Regular $n \times n$ Matrices with Elements in Z/p (p prime) 171

Appendix 3. Some Further Examples Illustrating the Standard Newton Method in $(Z/2)^n$ and $(Z/3)^n$ 173
Appendix 4. Continuous Iterations-Discrete Iterations 179

Bibliography 183

Index 193

Introduction

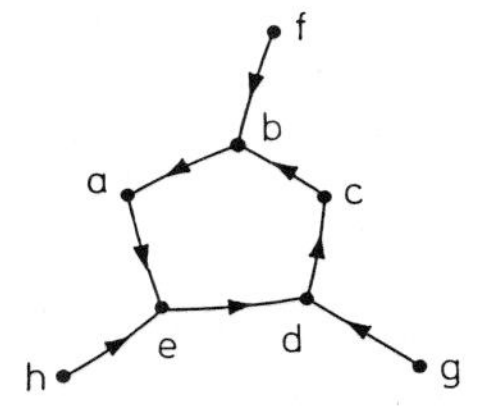

In presenting this monograph, I would like to indicate both its orientation as well as my personal reasons for being interested in *discrete* iterations (that is, iterations on a generally *very large, finite* set).

While working in numerical analysis I have been interested in two main aspects:

- *the algorithmic aspect*: an iterative algorithm is a mathematical entity which behaves in a *dynamic* fashion. Even if it is started far from a solution, it will often tend to get closer and closer.
- *the mathematical aspect:* this consists of a coherent and rigorous analysis of convergence, with the aid of mathematical tools (these tools are mainly the use of norms for convergence proofs, the use of matrix algebra and so on).

One may for example refer to the algorithmic and mathematical aspects of Newton's method in $\mathbb{R}^n$ as well as to the QR algorithm for eigenvalues of matrices. These two algorithms seem to me to be the most fascinating algorithms in numerical analysis, since both show a remarkable practical efficiency even though there exist relatively few global convergence results for them.

Both in teaching and research, and in particular with respect to the iterative solution of systems of equations, there are some facts which occur systematically:

1) The domain of practical use of the algorithms that are recognized as efficient amply exceed (luckily!) the theoretical domain in which convergence is proven mathematically. This may also be stated by saying that the (good) algorithms "work" much better than one is presently able to prove mathematically. Moreover, the usual sufficient conditions for convergence are often *conceptual*. This means that they are related to *intrinsic* prop-

erties of the problem, that are *not* accessible in numerical practice. From a theoretical point of view, these conditions are of course of interest. In practice, however, it is often not possible to test them. This situation often shocks students. They expect to be able to deal with actual numerical examples only with a complete knowledge of their theoretical structure. This is not realistic. Indeed, many good practical results are often obtained without knowing anything about convergence.

In fact, one may reasonably believe that the *mathematical analysis of convergence* may never completely reach *the reality of effective* convergence. *There seems to be something in the algorithmic dynamics that is outside the grasp of mathematical formalism.*

2) There are also cases where an iteration *does not* converge to a solution. Quite often in numerical analysis one simply gives sufficient conditions for convergence. Little thought is given to the case when these conditions are *not* satisfied. The problem of enlarging the convergence conditions and of investigating what happens when there is no convergence as well as the frequency of non-convergence is a difficult and largely open problem.

The idea, briefly, is that one should try to perform a better analysis for the intrinsic behaviour of an iteration, also when it does *not* converge.

3) The final point is, that to implement and run an iterative algorithm on a computer is no more than a *simulation* of the actual algorithm, using a *finite, discrete* tool. The algorithm was "theoretically" formulated and analysed in a "continuous" mathematical space (most often a normed vector space). It is, however, executed in a *discrete* space (due to the discrete nature of all digital devices). In such a space one can only *simulate* the defined algorithm. This leads to the usual problem of *numerical drift* resulting in the propagation of rounding errors.

In other words: the analysis and teaching of the convergence of a numerical algorithm usually takes place in a normed vector space ($\|x-x^r\|\to 0$), resulting in a limit type approach to the mathematical solution. This is rather *paradoxical* since the real machine convergence is a *stationary* convergence where for example, 4, 6, or 8 decimal digits are "clearly" stabilized at the end of a finite number of iterations.

One may of course still claim that, in spite of numerous attempts to clarify the "machine" concepts, one only knows that there is a jungle of numbers and operations in a real machine, together with a complicated structure. Moreover this structure is variable from one machine to another.

It is furthermore clear that a mathematical analysis of the problem posed and of the algorithm considered is vital for a fundamental and coherent understanding of their structure.

Moreover it appears that machine results frequently (surprisingly often) are much better than one might expect. Indeed, the good quality of the practical results generally obtained on a computer is really a bit surprising, considering the above remarks!

At this point of these rather classical reflections on numerical analysis, I would like to mention the fact that, over a number of years, *discrete iterative models* have been developed by many people in diverse fields.

With this in mind, we started a working group on discrete and continuous iterations in Grenoble some years ago. This group consists of Michel Cosnard, André Eberhard, Eric Goles, Yves Robert, Maurice Tchuente, myself, as well as a number of young researchers who worked with us at various times while writing a thesis: Jean-Christian Angles d'Auriac, Souad El Bernoussi, Mariano Bona, Zequ Jiang, Marc Legendre, Christophe Masse, Lamine Melkemi, Jean-Michel Muller, Xin An Pan, Didier Pellegrin, Hans Detlef Schulz, Houssine Snoussi, Denis Trystram ... We were able to distinguish the following fields, where *discrete* iterative models were being developed:

Theoretical Computer Science: Transition systems, Petri nets, synchronisation, program proving, distributed control, automata networks, cellular automata, parallelism, systolic arrays ...

Biomathematics: Discrete models for interactions between neurons (neural networks), between genes (genetic nets), booleean networks, threshold automata ...

Physics: Discrete models for disordered matter, in particular for the spin glass problem, frustration, percolation, phase transitions ...

Sociology, Psychology: Discrete models for interactions, between people, between populations ... (the bibliography at the end of this book is an attempt to make an inventory of these different fields of research).

The *analysis* of the iterative behaviour of these *discrete* models is only in its infancy. Indeed quite often one *simulates* and one tries to *observe* regular behaviours, but one doesn't know exactly how to analyse these behaviours mathematically. For this, one needs *new tools* and *new concepts*. One feels that this subject is therefore an attractive research area. In the different fields quoted above, papers appear, often at the interface between several disciplines: Biology, Physics, Computer science, Mathematics, and so on.

Recently there have been a number of conferences on the unifying theme of iterations and of automata networks. During these conferences, researchers from *different* areas could meet, whose different interests were directly connected to the theme. This resulted in fruitful interactions (interactions between people at a conference on automata networks is a good example of an automata network ...) New links appeared, and joint work was developed. Further conferences have also been planned on these topics; the scientific life is vibrating ...

Often, for the study of the iterative behaviour of such automata networks, or more generally, for discrete models, the existing studies are of an algebraïc kind. This is quite natural, in the context discussed here.

The idea and the aim of this monograph is therefore to attempt to *analyse* the behaviour of iterations in a discrete setting. For this, we *use the notion of a stationary convergence arising from the use of a metric tool.* (This tool is introduced so that one is able to speak about *the distance between two configurations.* In fact, it is a *boolean, vectorial distance* that turned out to be the more suitable for our purposes.) With the aid of this tool, one tries to transpose the usual convergence results for iterations in normed vector spaces into *discrete* spaces. (These are classical *global* or *local* results, that have withstood the test of time).

The aim of this monograph is indeed to show that this transposition may be carried "rather far".

The results comprise contraction, attractive cycles and fixed points, comparisons of operator modes (parallel, serial-parallel, serial), Newton's method and so on.

I would like to state at this point that this study and transposition interested me very much since 1973: the context of the applications which I described above has especially served as a reference point for me.

Even though I have concentrated on proving results analogous to results in numerical analysis and linear algebra, I would also like to point out that this is not the only possible approach. Other approaches exist in different contexts. (I am in particular thinking of the approaches taken in Cellular Automata, in Computer Science, in Physics and in Biomathematics.)

Indeed, there are a number of things that might be elaborated. In particular one of the main "hot" topics, both in computer science as well as in biomathematics, is the study of *asynchronous iterations:* The asynchronous iterations in Numerical Analysis go back at least to the 1969 fundamental work by D. Chazan, W. Miranker about the (so-called) chaotic relaxation. Ref. [21].

This monograph only deals with synchronous iterations. It is, however, clearly possible to study asynchronous iterations as well, in the "discrete metric" context of this work.

As a final comment we note that in the context of this monograph, the point 3 above (numerical drift) does not apply since the numerical calculations follow the theory exactly. Point 1 is still valid and for point 2 it should be noted that a study of cycles remains to be developed.

Most of the contents of this monograph appeared as a sequence of papers [112] to [117]. Although they have been rewritten for the sake of consistency, as well as translated into English, the various chapters are essentially found as:

Chapter 3: Ref. [112]

Chapter 4: Ref. [113]

Chapter 6: Ref. [114]

Chapter 7: Ref. [115]

1. Discrete Iterations and Automata Networks: Basic Concepts

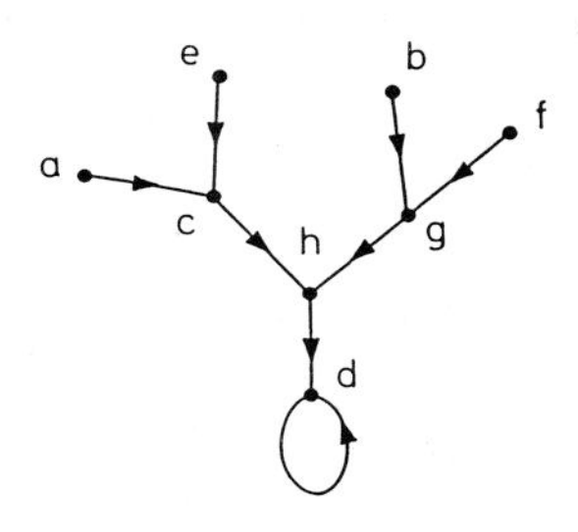

1. Discrete Iterations and Their Graphs

In the sequel X will always denote a finite, generally very large set*, and F will denote a map of X into itself.

We are interested in the following iteration scheme: *starting with an* x^0 *from* X, *the sequence of successive approximations to* F is defined by

$$x^{r+1} = F(x^r) \qquad (r = 0, 1, 2, \ldots).$$

This iteration scheme is called a *discrete iteration* since X is finite.

The main question of interest consists of investigating the behaviour of this sequence, given particular assumptions for F.

Since X is finite, it is clear that the iteration defined above has only two possible modes of behaviour**.

– It is possible that the sequence generated by this iteration will remain stationary at an element ξ possibly after some initial steps. In this case the iteration is said to *converge* to ξ. Clearly it follows immediately that ξ is a *fixed point* for F since $F(\xi) = \xi$.

– If the sequence does not converge in the above sense, then it follows that it must repeat itself after a certain number of steps since X is finite.

---

* *Example (typical)*. Consider a grid consisting of $p \times p$ nodes, each node being associated with a value 0 or 1. The set X is taken to be the set of all possible 0 or 1 configurations on this grid and the cardinality of this set is 2^{p^2}. (Even for $p=5$, $2^{p^2} = 2^{25} = 33554432$; for $p=10$ one obtains $2^{p^2} = 2^{100}$ which is of the order 10^{30}.)

** For a discussion of the differences between this case and the case where F maps a *continuous* set into itself (for example the interval [0, 1]) see [1]–[14]. In the latter case matters may be *much more complex* than they are here.

The elements that are repeated in the iteration sequence constitute a *cycle* $\{\xi_1, \xi_2, \ldots, \xi_p\}$ and this cycle is defined by

$$\begin{aligned} \xi_2 &= F(\xi_1) \\ &\vdots \\ \xi_p &= F(\xi_{p-1}) \\ \xi_1 &= F(\xi_p). \end{aligned} \qquad (\xi_1 \ldots \xi_p \text{ are pairwise distinct}).$$

This cycle is said to have *length* p and it should be noted that a fixed point is nothing but a cycle of length 1.

Although the complete tale has now been told (!!!) it seems appropriate to describe the behaviour in more detail.

Definitions. The *iteration graph* for F is the graph consisting of vertices which are elements of X and the following arcs: for all x in X, an arc connects x to $F(x)$.

If x and z are in X, one says that z is a *descendant* of x if there exists an integer $p \geq 0$ (p may be zero) such that $z = F^p(x)$. (All x in X are their own proper descendants since $p=0$ is allowed.)

Two elements u and v in X are said to be *equivalent* if they have a common descendant.

It is easy to see that this relation is truly an equivalence relation on X which partitions X into a finite number of equivalence classes of finite cardinality since X is finite. These equivalence classes in turn divide the graph of F into mutually disjoint, but connected sub-graphs called the *connected components* of the graph or the *basins* of F.

Theorem 1. *Let C be a basin in the partition defined above. Then either*

– *C contains a fixed point ξ of F which is unique in C. Furthermore for all x^0 in C, the iteration $x^{r+1} = F(x^r)$ remains in C and converges to ξ.*

or

– *C does not contain a fixed point. Therefore for all x^0 in C, the iteration $x^{r+1} = F(x^r)$ remains in C and runs through the same cycle indefinitely after possibly some transient iteration steps.*

The proof is elementary and goes as follows:

– Suppose C contains two fixed points ξ and η. From the definition of C they have a common descendant. Since ξ and η both are fixed points it follows that $\xi = \eta$ from which we conclude that if C has a fixed point, then it is unique.

For all x^0 it follows from the definition of C that x^0 and ξ are equivalent, which in turn implies the existence of integers s and q such that

$$F^s(x^0) = F^q(\xi) = \xi.$$

Consequently, if the iteration $x^{r+1}=F(x^r)$ is started at x^0, then it is stationary at ξ if $r \geq s$, that is

$$x^r = \xi \qquad \forall r \geq s.$$

– Suppose now that the basin C does not contain a fixed point. Starting with an x^0 in C it is then clear that the iteration $x^{r+1}=F(x^r)$ $(r=0, 1, 2, \ldots)$ remains in C and that it does not converge. This means that since C is finite the iteration must run through the *same* cycle indefinitely after possibly some transient iteration steps. □

2. Examples

1) $X=\{0, 1, 2, \ldots, 9\}$.

The iteration F is defined in the following manner: for a in X, a is squared and then the sum of the digits of a is formed. If this sum is greater than 9, then the sum of the digits of this sum is formed once more such that in every case an integer between 0 and 9 is obtained.

The above iteration results in the following iteration graph:

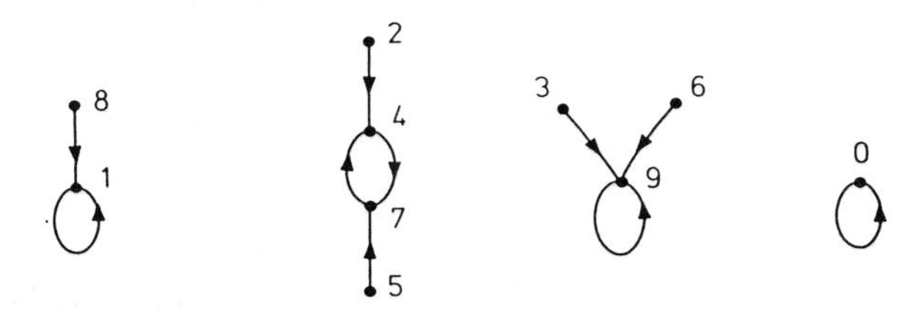

There are four basins, the second one contains a cycle.

2) $X=\{0, 1\}^6$ is identified with $(Z/2)^6$ (64 elements)

(Z is the integers, $Z/2 \equiv Z \bmod 2$). The following affine transformation is defined:

$$x \in (Z/2)^6 \rightarrow F(x) = Ax + b$$

(operations in $Z/2$: $1+1=0$)

with

$$A=\begin{bmatrix} 0 & 0 & 1 & 1 & 0 & 0 \\ 0 & 0 & 0 & 0 & 1 & 1 \\ 1 & 0 & 1 & 1 & 0 & 0 \\ 0 & 0 & 1 & 0 & 0 & 0 \\ 0 & 1 & 0 & 0 & 0 & 0 \\ 0 & 1 & 0 & 0 & 0 & 0 \end{bmatrix} \quad \text{and} \quad b=\begin{bmatrix} 0 \\ 1 \\ 1 \\ 0 \\ 0 \\ 0 \end{bmatrix}.$$

The graph of the iteration for F consists of two basins one having a fixed point and the other a cycle of length 7. (Displayed in this manner the pattern has a remarkable regularity.)

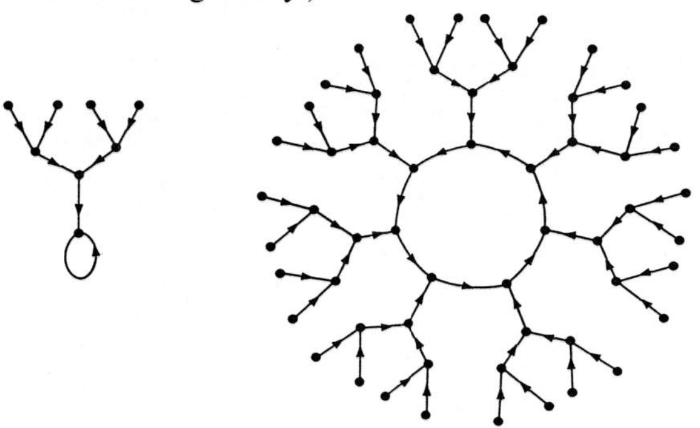

3) In the following examples $X=\{0,1\}^n$. We define a linear map

$$F(x)=Ax$$

where A is an $n\times n$ matrix with elements from $\{0,1\}$ and where the operations are *boolean* operations, that is $1+1=1$ etc.

a) $n=6$

$$A=\begin{bmatrix} 1 & 1 & 0 & 0 & 0 & 0 \\ 1 & 0 & 1 & 0 & 0 & 0 \\ 0 & 1 & 0 & 1 & 0 & 0 \\ 0 & 0 & 1 & 0 & 1 & 0 \\ 0 & 0 & 0 & 0 & 0 & 1 \\ 1 & 0 & 0 & 0 & 0 & 0 \end{bmatrix}.$$

This iteration has the following iteration graph:

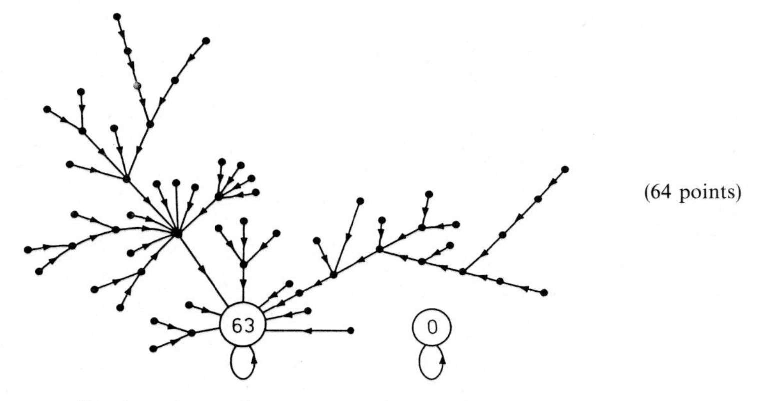

(64 points)

There are two fixed points, the zero vector and the vector having all elements equal to 1 (indicated on the graph by their decimal equivalents 0 and 63).

b) $n=5$

$$\begin{bmatrix} 1 & 1 & 0 & 0 & 0 \\ 0 & 0 & 1 & 0 & 0 \\ 0 & 0 & 0 & 1 & 0 \\ 0 & 0 & 0 & 0 & 1 \\ 1 & 0 & 0 & 0 & 0 \end{bmatrix}.$$

The iteration graph for F is

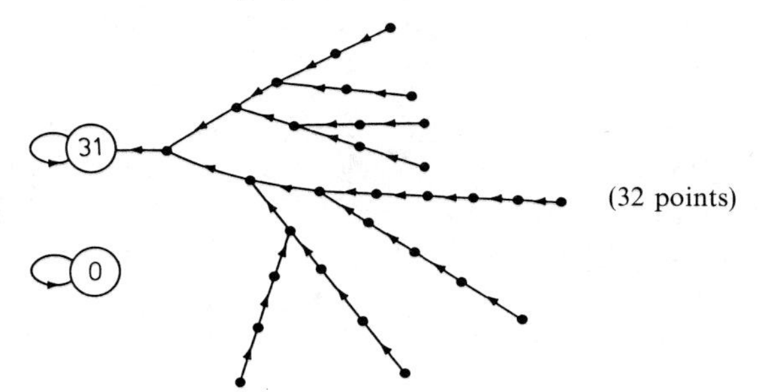

(32 points)

c) $n=8$

$$A=\begin{bmatrix} 0 & 1 & 0 & 0 & 0 & 0 & 0 & 0 \\ 0 & 0 & 1 & 0 & 0 & 0 & 1 & 0 \\ 0 & 0 & 0 & 1 & 0 & 0 & 0 & 0 \\ 0 & 0 & 0 & 0 & 1 & 0 & 0 & 0 \\ 0 & 0 & 0 & 0 & 0 & 1 & 0 & 0 \\ 0 & 0 & 1 & 0 & 0 & 0 & 1 & 0 \\ 0 & 0 & 0 & 0 & 0 & 0 & 0 & 1 \\ 1 & 0 & 0 & 0 & 0 & 0 & 0 & 0 \end{bmatrix}.$$

The iteration graph for F is

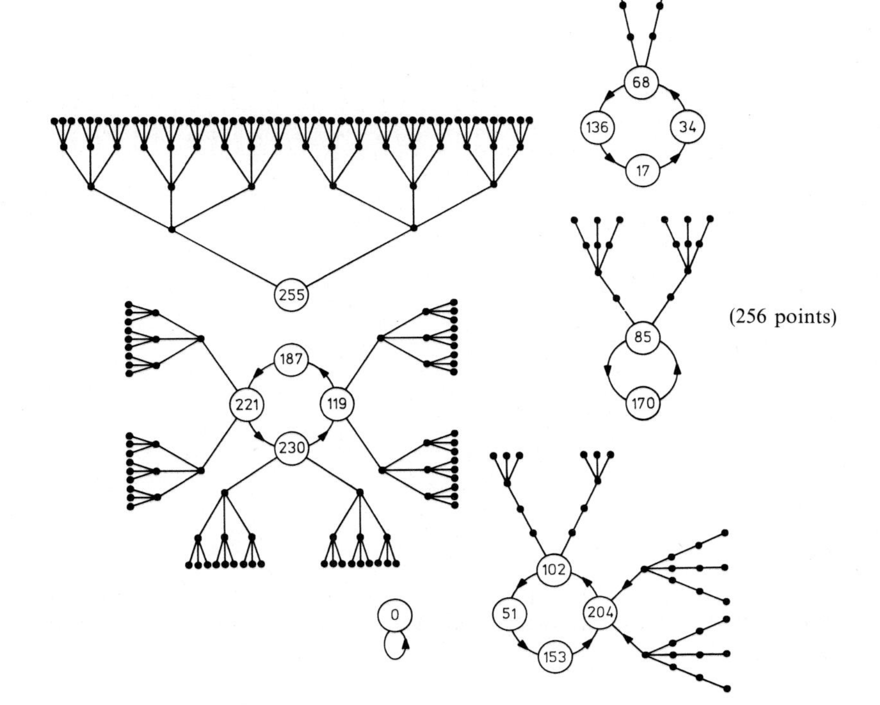

(256 points)

It consists of 6 basins, 2 fixed points (the zero vector and the vector having all components equal to 1), 3 cycles of length 4 and 1 cycle of length 2.

Remarks. (1) In order to draw the iteration graph for F it is necessary to calculate the map of every member of the finite set X by F and then to "link up the pieces" in order to finally produce a convenient graphical layout. One easily realizes that this could be a complicated process when X contains many elements.

(2) If $x^{r+1}=F(x^r)$ converges to the same fixed point ξ independently of x^0 *then the iteration graph has only one basin and this basin has a unique fixed point*. In this case we say that F (or its iteration graph) is *simple*. Here is an example

$$X=\{0,1\}^7=(Z/2)^7 \quad \text{contains 128 elements,}$$
$$F(x)=Ax \qquad \text{(Linear transformation in } (Z/2)^7)$$

with

$$A=\begin{bmatrix} 1 & 0 & 0 & 0 & 1 & 0 & 0 \\ 0 & 1 & 1 & 0 & 0 & 0 & 0 \\ 1 & 1 & 0 & 0 & 0 & 0 & 0 \\ 0 & 0 & 1 & 0 & 0 & 0 & 0 \\ 0 & 1 & 1 & 0 & 0 & 0 & 0 \\ 0 & 0 & 0 & 1 & 1 & 0 & 0 \\ 0 & 1 & 1 & 0 & 1 & 1 & 0 \end{bmatrix}.$$

The iteration graph for F is simple

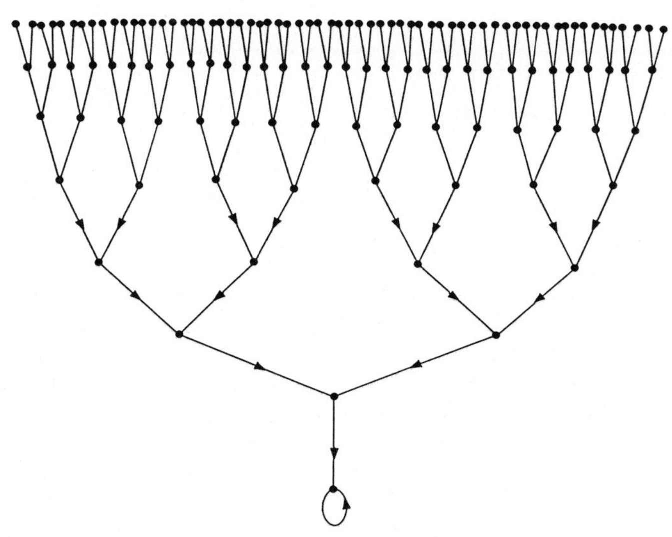

and the pattern is again very regular.

In the remainder of this study we establish certain results for the iteration graph for F starting with some conditions on F.

We will now assume that X is the cartesian product of a finite number of finite sets in order to obtain these kinds of results.

3. Connectivity Graphs and Incidence Matrices

In the sequel one is given n *finite* sets $X_1, X_2, \ldots, X_n$ and their cartesian product

$$X = \prod_{i=1}^{n} X_i.$$

An element x of X is therefore an n-tuple $x=(x_1, \ldots, x_n)$ where $x_i \in X_i$ $(i=1,2,\ldots,n)$. The most common examples (see the Introduction as well as the above examples) are given by $X_i=\{0,1\}$. Therefore let $X=\{0,1\}^n$ and one may identify X with the set of the 2^n vertices of the n-cube. This is the usual way of looking at X.

A mapping F of X into itself may also be formalized by writing the relation $y=F(x)$ as

$$\begin{cases} y_1 = f_1(x_1, \ldots, x_n) \\ \vdots \\ y_i = f_i(x_1, \ldots, x_n) \\ \vdots \\ y_n = f_n(x_1, \ldots, x_n) \end{cases}$$

where the f_i are the *components* of F.

Definitions. The *incidence matrix* of F is the $n \times n$ boolean matrix

$$B(F) = (b_{ij}) \quad \text{where} \quad \begin{cases} b_{ij}=0 & \text{if } f_i \text{ does not depend on } x_j \\ b_{ij}=1 & \text{otherwise.} \end{cases}$$

The *connectivity graph* of F is the graph containing the vertices $P_1, P_2, \ldots, P_n$ linked as follows:

an arc connects P_j to P_i if f_i depends on x_j, that is to say $b_{ij}=1$.

Clearly the connectivity graph and the incidence matrix both provide the same information for F. This information is fairly coarse, it is of the type all or nothing (f_i depends or does not depend on x_j and no further information). *We now proceed to exploit this type of information for F in order to study its iteration graph* (see for example Chaps. 2 and 4).

Remark. In the most common case, namely where $X=\{0,1\}^n$, one may visualize the iteration graph for F as a map of the n-cube into itself. The *connectivity graph* for F then has n vertices, whereas the *iteration graph* for F has 2^n vertices.

Example. $n=3$ $X_i=\{0,1\}$ $(i=1,2,3)$ $X=\{0,1\}^3$.

Let F be defined by the relation $y=F(x)$ (using boolean notation):

$$\begin{aligned} y_1 &= x_1\bar{x}_2+x_3 = f_1(x_1,x_2,x_3)\\ y_2 &= x_1+\bar{x}_3 \quad = f_2(x_1,\sqcup,x_3)^{*}\\ y_3 &= x_2x_3 \quad\quad = f_3(\sqcup,x_2,x_3). \end{aligned}$$

The *incidence matrix* for F and its *connectivity graph* are now

$$B(F)=\begin{bmatrix}1&1&1\\1&0&1\\0&1&1\end{bmatrix}$$

P1 P2 P3

($n=3$ vertices)

We can furthermore give the tabulation for F and its *iteration graph* as

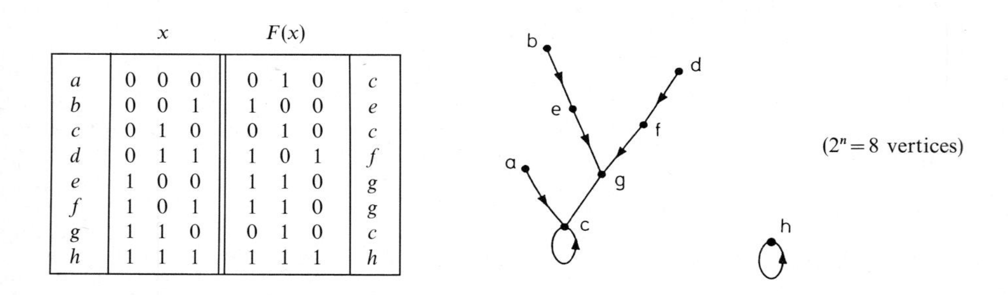

	x			$F(x)$			
a	0	0	0	0	1	0	*c*
b	0	0	1	1	0	0	*e*
c	0	1	0	0	1	0	*c*
d	0	1	1	1	0	1	*f*
e	1	0	0	1	1	0	*g*
f	1	0	1	1	1	0	*g*
g	1	1	0	0	1	0	*c*
h	1	1	1	1	1	1	*h*

($2^n=8$ vertices)

The last graph may be visualized in the 3-cube as:

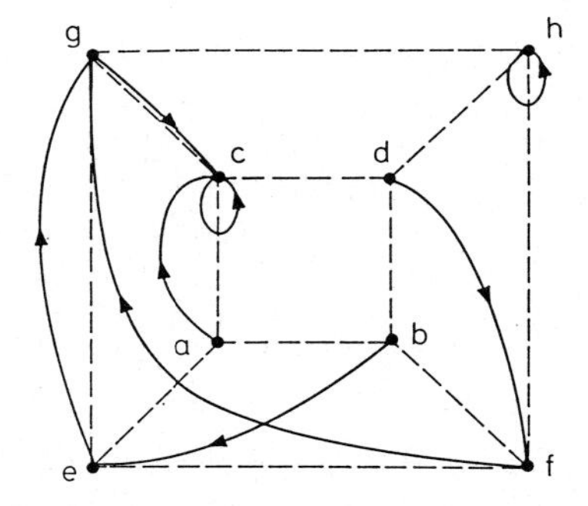

* The notation $\sqcup$ means that f_2 does not depend on x_2 in this case.

4. Interpretations in Terms of Automata Networks (see [63]–[141])

One may use the language of automata networks in order to describe the preceding notions.

Therefore let $X = \prod_{i=1}^{n} X_i$ (X_i finite) and $F: X \to X$.

Consider then the *connectivity graph* for F. At each node i (vertex i) of this graph, we define a *finite state automaton* (also called a *cell*) whose states are from the finite set X_i.

An element $x = (x_1, \ldots, x_n)$ from X is visualized as a *configuration* of the network, that is each automaton i is in the state $x_i \in X_i$ ($i = 1, 2, \ldots, n$).

The automaton placed at each node i of the network is defined by the *transition function* f_i, that is it depends only on the variables corresponding to vertices j having an arrow to vertex i. In this manner F defines a network of automata, and it is clear that the *connectivity graph* for F is nothing but the *connectivity graph* for the network*.

Now, starting from an *initial configuration* x^0, the method of *successive approximations* to F defined by

$$x^{r+1} = F(x^r) \qquad (r = 0, 1, 2, \ldots)$$

consists of *the parallel operation of the network.* At each discrete clock pulse, each automaton takes cognizance of its environment and then evaluates the transition function f_i (that is, *adapts*) according to

$$x_i^{r+1} = f_i(x_1^r, \ldots, x_n^r) \qquad (i = 1, 2, \ldots, n).$$

Remarks. (1) A *stable configuration* (with respect to F) is a configuration of the network that remains invariant during one step of the iteration, that is, it is a *fixed point* $\xi = (\xi_1, \ldots, \xi_n)$ of F with

$$\xi_i = f_i(\xi_1, \ldots, \xi_n) \qquad (i = 1, 2, \ldots, n) \quad \text{i.e. } \xi = F(\xi).$$

(2) For the sake of convenience, one sometimes uses the language of iterations (r-th iterate, fixed point, etc.) and sometimes the language of automata networks (configuration, transition function, etc.).

(3) The interpretation in terms of automata networks is of interest whenever the incidence matrix is relatively *sparse*, that is, whenever the connectivity graph is relatively disconnected. When the connectivity graph is completely connected (all the vertices are joined together by arcs) then the incidence matrix is full of 1's and it will therefore carry no precise information.

---

* It is sometimes said that the f_i's are the *local transition functions* and that F is the *global transition function* (with the obvious interpretations).

Example. Consider the following (very rudimentary!) model of a neuron in the human brain:

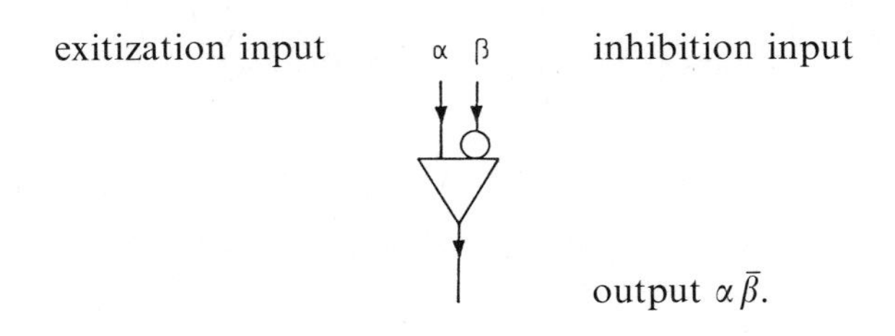

From the boolean input variables α and β the neuron generates the boolean function $\alpha\bar{\beta}$.

Let us now construct a network of 3 neurons

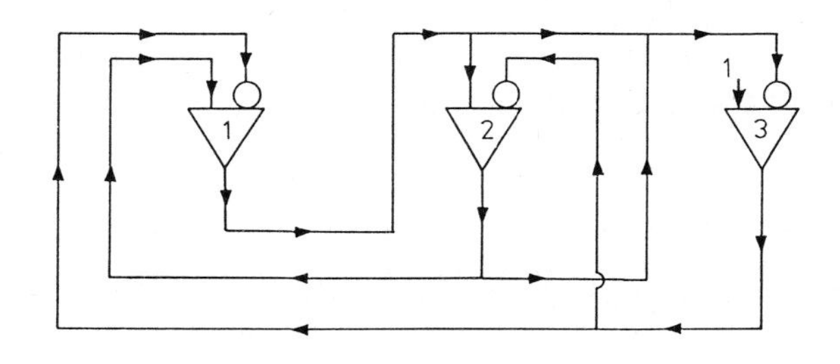

with the conventions

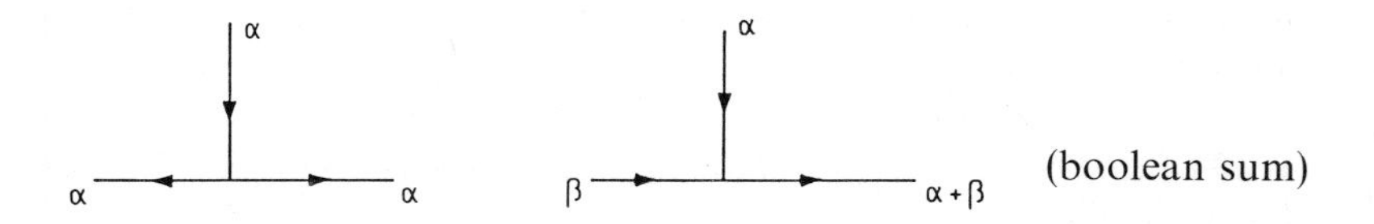

Each of the neurons may be assimilated into a cell that may take on the 2 binary states, 0 and 1. The state of the cell is by definition the boolean value of the output of the neuron.

One therefore has $X_1 = X_2 = X_3 = \{0, 1\}$ and $X = \{0, 1\}^3$.

It is evident that F is defined in the following manner (using boolean notation):

$$f_1(x_1, x_2, x_3) = x_2\bar{x}_3 = f_1(\sqcup, x_2, x_3)$$
$$f_2(x_1, x_2, x_3) = x_1\bar{x}_3 = f_2(x_1, \sqcup, x_3)$$
$$f_3(x_1, x_2, x_3) = 1 \cdot \overline{x_1 + x_2} = \bar{x}_1\bar{x}_2 = f_3(x_1, x_2, \sqcup).$$

The incidence matrix and the connectivity graph are given by

$$B(F)=\begin{bmatrix}0 & 1 & 1\\ 1 & 0 & 1\\ 1 & 1 & 0\end{bmatrix}$$

The iteration graph for F (corresponding to a *parallel* operation of the automata) is given by

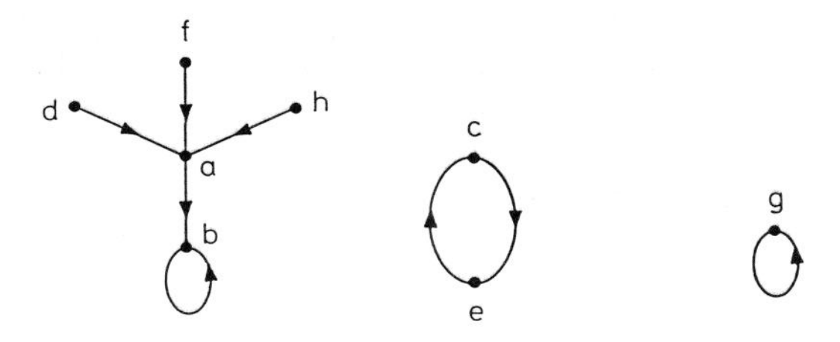

It consists of 3 basins, 2 fixed points (stable configurations) and a cycle of length 2.

5. Serial Operation and the Gauss-Seidel Operator

The iteration:

$$x^{r+1}=F(x^r) \qquad (r=0,1,2,\ldots)$$

corresponds to a *purely parallel mode of operation* of the associated automata network. Indeed at each discrete clock pulse (indicated by r) all cells adapt *simultaneously*.

We will now define a set of operating modes called *serial-parallel operating modes*. For this it is assumed that the cells have a fixed numbering $(P_1, P_2, \ldots, P_n)$.

A *purely serial mode of operation* is first described. Starting from a configuration x^r, the *first cell* adapts first, then the *second cell* adapts (taking into account the effects of the potential changes in the first cell) and so on. This results in the scheme

$$\begin{aligned}
x_1^{r+1}&=f_1(x_1^r, x_2^r, \ldots, x_n^r)\\
x_2^{r+1}&=f_2(x_1^{r+1}, x_2^r, \ldots, x_n^r)\\
&\vdots\\
x_n^{r+1}&=f_n(x_1^{r+1}, x_2^{r+1}, \ldots, x_{n-1}^{r+1}, x_n^r) \qquad (r=0,1,2,\ldots)
\end{aligned}$$

which is nothing but *the Gauss-Seidel iteration for F*.

In the sequel, F_i is always the operator defined by

$$x=(x_1 \dots x_n) \to F_i(x) = \begin{bmatrix} x_1 \\ \vdots \\ f_i(x_1 \dots x_n) \\ \vdots \\ x_n \end{bmatrix}.$$

This is the operator that with a given configuration only adapts the i-th cell (according to the transition function f_i). One therefore has:

Theorem 2. *The Gauss-Seidel method is effectively a method of successive approximations to an operator G, called the associated Gauss-Seidel operator for F.*

$$G = F_n \circ \dots \circ F_2 \circ F_1.$$

F and G have the same set (possibly empty) of fixed points (stable configurations).

To show this, the following relations are written for all $x=(x_1, \dots, x_n)$ from X:

$$\begin{aligned} g_1(x_1, \dots, x_n) &= f_1(x_1, \dots, x_n) \in X_1 \\ g_2(x_1, \dots, x_n) &= f_2(g_1(x), x_2, \dots, x_n) \in X_2 \\ &\vdots \\ g_i(x_1, \dots, x_n) &= f_i(g_1(x), \dots, g_{i-1}(x), x_i, \dots, x_n) \in X_i \\ &\vdots \\ g_n(x_1, \dots, x_n) &= f_n(g_1(x), \dots, g_{n-1}(x), x_n) \in X_n. \end{aligned}$$

In this manner an operator G is defined from X into itself and the conclusions of the theorem should be obvious since the Gauss-Seidel iteration may be written

$$x^{r+1} = G(x^r) \qquad (r=0, 1, 2, \dots)$$

where the g_i are the components of G. □

Examples. (1) Let us calculate G for the example on p.8

$$X = \{0, 1\}^3$$

$$\begin{aligned} f_1(x) &= x_1 \bar{x}_2 + x_3 & & g_1(x) = x_1 \bar{x}_2 + x_3 \\ f_2(x) &= x_1 + \bar{x}_3 & \text{from which } & g_2(x) = x_1 \bar{x}_2 + x_3 + \bar{x}_3 = 1 \\ f_3(x) &= x_2 x_3 & & g_3(x) = 1 \cdot x_3 = x_3. \end{aligned}$$

(boolean notation)

This implies that

$$B(F) = \begin{bmatrix} 1 & 1 & 1 \\ 1 & 0 & 1 \\ 0 & 1 & 1 \end{bmatrix}, \qquad B(G) = \begin{bmatrix} 1 & 1 & 1 \\ 0 & 0 & 0 \\ 0 & 0 & 1 \end{bmatrix}.$$

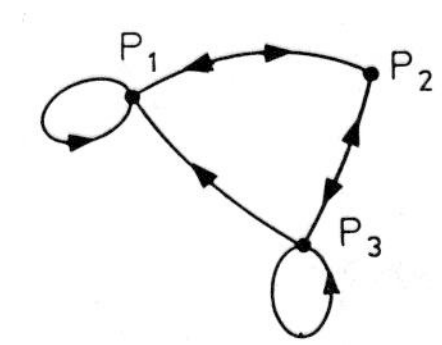

Connectivity graph for F

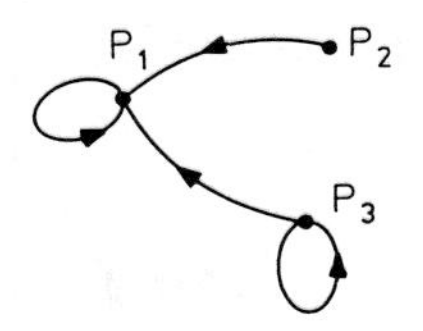

Connectivity graph for G

Tables for F and G

	x			$F(x)$				$G(x)$			
a	0	0	0	0	1	0	c	0	1	0	c
b	0	0	1	1	0	0	e	1	1	1	h
c	0	1	0	0	1	0	c	0	1	0	c
d	0	1	1	1	0	1	f	1	1	1	h
e	1	0	0	1	1	0	g	1	1	0	g
f	1	0	1	1	1	0	g	1	1	1	h
g	1	1	0	0	1	0	c	0	1	0	c
h	1	1	1	1	1	1	h	1	1	1	h

One may calculate the table for G starting from the table for F in a general manner without going back to the analytical expression for f_i and g_i. In fact, for this example, one may read directly from the table for F that

$$g_1(b)=f_1(b)=1$$
$$g_2(b)=f_2(1\ 0\ 1)=1$$
$$g_3(b)=f_3(1\ 1\ 1)=1 \quad \text{from which} \quad G(b)=(1\ 1\ 1)=h.$$

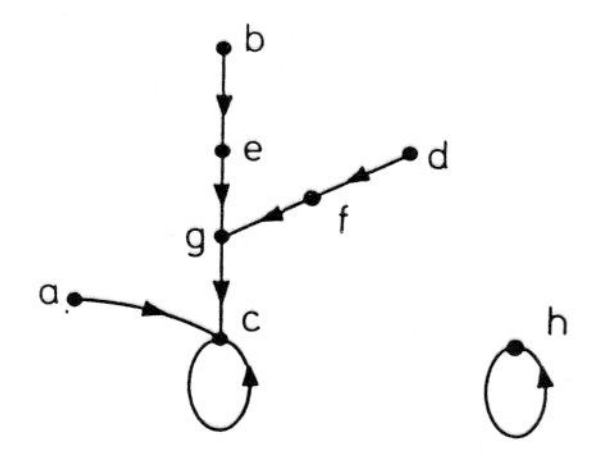

Iteration graph for F

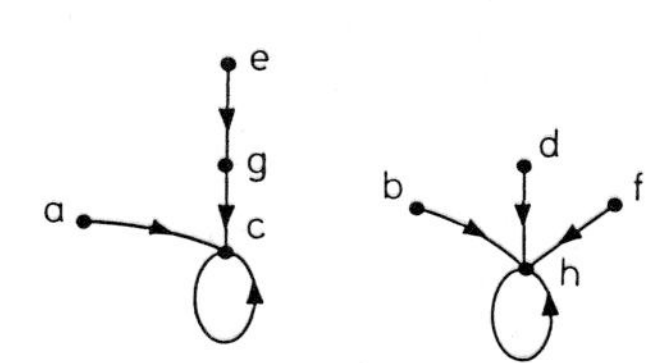

Iteration graph for G

Clearly, both F and G have the same fixed points.

(2) $X=\{0,1\}^5$ $F(x)=Ax$ (boolean operations) with

$$A=\begin{bmatrix} 0 & 1 & 0 & 0 & 0 \\ 0 & 0 & 1 & 0 & 0 \\ 0 & 0 & 0 & 1 & 0 \\ 0 & 0 & 0 & 0 & 1 \\ 1 & 0 & 1 & 0 & 0 \end{bmatrix}=B(F).$$

Then $G(x) = Mx$ with

$$M = \begin{bmatrix} 0 & 1 & 0 & 0 & 0 \\ 0 & 0 & 1 & 0 & 0 \\ 0 & 0 & 0 & 1 & 0 \\ 0 & 0 & 0 & 0 & 1 \\ 0 & 1 & 0 & 1 & 0 \end{bmatrix} = B(G)$$

and we obtain

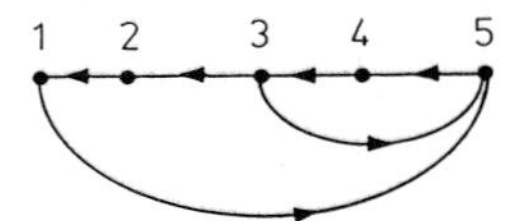

Connectivity graph for F

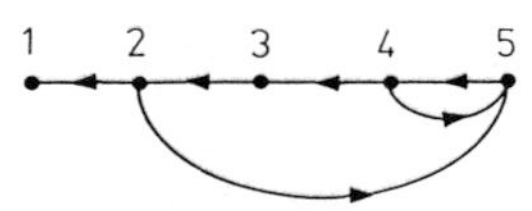

Connectivity graph for G

(*Remark*. In this example $B(F)$ and $B(G)$ are not comparable. In particular *we do not have* $B(G) \leq B(F)$ (the elementwise partial order relation based on $0 \leq 0 \leq 1 \leq 1$ in $\{0, 1\}$)).

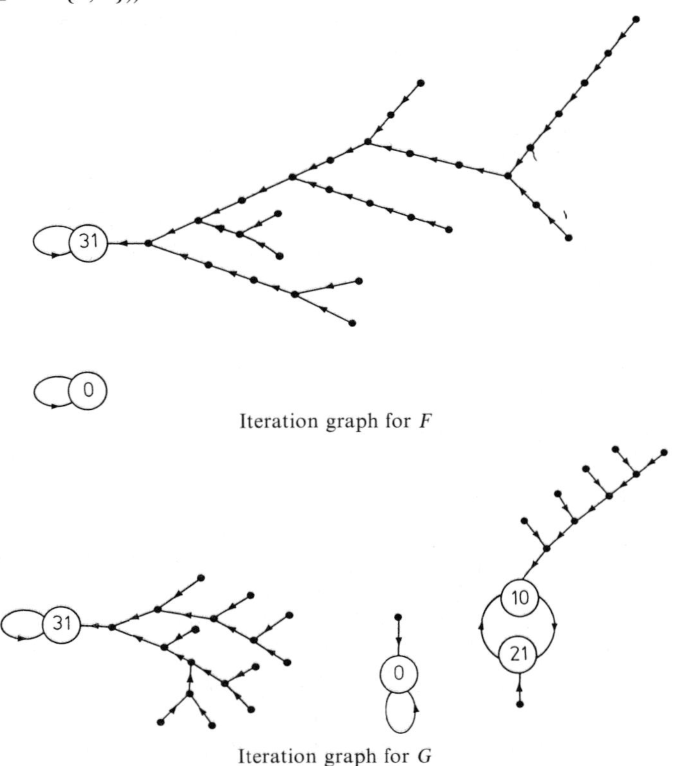

Iteration graph for F

Iteration graph for G

The iteration graphs for F and G have the same fixed points. The iteration graph for G contains a basin with a cycle (of length 2) which is not present in the iteration graph for F, however.

(3) $X=\{0,1\}^4=\{a,b,\ldots,p\}$ F is defined by:

$$\begin{array}{lcl} f_1(x)=x_4 & & g_1(x)=x_4 \\ f_2(x)=x_1+\bar{x}_4 & \text{from which} & g_2(x)=x_4+\bar{x}_4=1 \\ f_3(x)=\bar{x}_2+x_1 & & g_3(x)=x_4 \\ f_4(x)=\bar{x}_3 & & g_4(x)=\bar{x}_4 \end{array}$$

and where

$$B(F)=\begin{bmatrix} 0 & 0 & 0 & 1 \\ 1 & 0 & 0 & 1 \\ 1 & 1 & 0 & 0 \\ 0 & 0 & 1 & 0 \end{bmatrix}, \qquad B(G)=\begin{bmatrix} 0 & 0 & 0 & 1 \\ 0 & 0 & 0 & 0 \\ 0 & 0 & 0 & 1 \\ 0 & 0 & 0 & 1 \end{bmatrix}.$$

Connectivity graphs

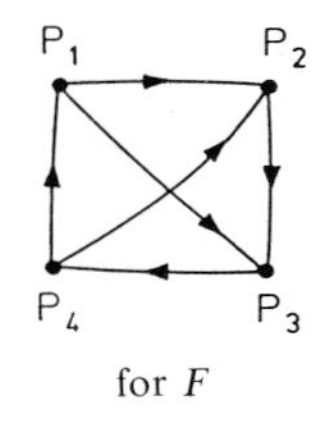

for F

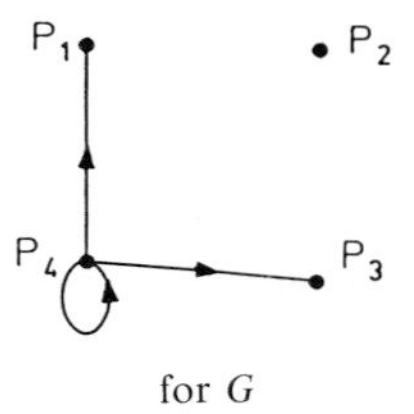

for G

Tables for F and G

	x	$F(x)$		$G(x)$	
a	0 0 0 0	0 1 1 1	h	0 1 0 1	f
b	0 0 0 1	1 0 1 1	l	1 1 1 0	o
c	0 0 1 0	0 1 1 0	g	0 1 0 1	f
d	0 0 1 1	1 0 1 0	k	1 1 1 0	o
e	0 1 0 0	0 1 0 1	f	0 1 0 1	f
f	0 1 0 1	1 0 0 1	j	1 1 1 0	o
g	0 1 1 0	0 1 0 0	e	0 1 0 1	f
h	0 1 1 1	1 0 0 0	i	1 1 1 0	o
i	1 0 0 0	0 1 1 1	h	0 1 0 1	f
j	1 0 0 1	1 1 1 1	p	1 1 1 0	o
k	1 0 1 0	0 1 1 0	g	0 1 0 1	f
l	1 0 1 1	1 1 1 0	o	1 1 1 0	o
m	1 1 0 0	0 1 1 1	h	0 1 0 1	f
n	1 1 0 1	1 1 1 1	p	1 1 1 0	o
o	1 1 1 0	0 1 1 0	g	0 1 0 1	f
p	1 1 1 1	1 1 1 0	o	1 1 1 0	o

Finally, the iteration graphs for F and G are

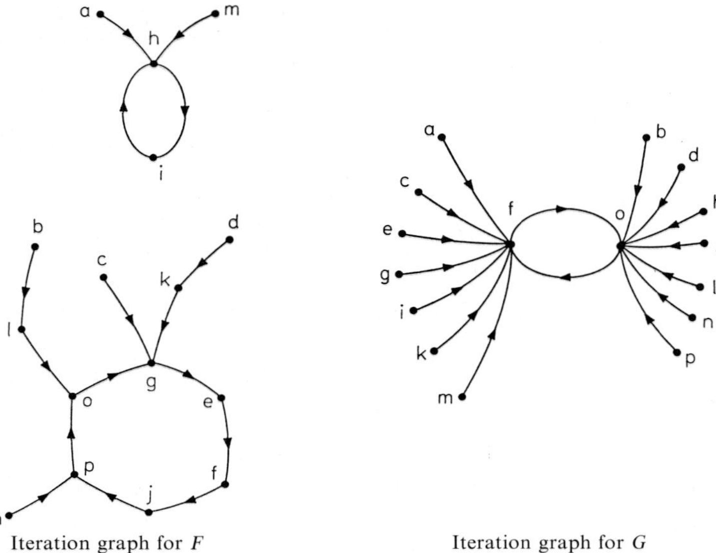

Iteration graph for F Iteration graph for G

Neither F nor G have any fixed points. Both have basins with cycles in their iteration graphs. The graph for G, however, is much "more regular" than the graph for F.

Remarks. (1) It is clear that the incidence matrix $B(F_i)$ for the operator F_i is nothing but the boolean matrix which has the i-th line copied from $B(F)$ and the others from the unit matrix giving

$$B(F_i)=\begin{bmatrix} 1 & & & & & & \\ & 1 & & & & & \\ & & \ddots & & & & \\ & & & 1 & & & \\ b_{i1} & \cdots & & b_{ii} & \cdots & & b_{in} \\ & & & & 1 & & \\ & & & & & \ddots & \\ & & & & & & 1 \end{bmatrix} \quad \text{(elements not shown are zero).}$$

(2) The previous theorem says that the serial mode of operation for the automata network defined by F is identical to the parallel mode of operation for the automata network defined by G.

It is therefore natural to call the iteration graph for G the *serial mode iteration graph.* These graphs are of course a priori different. They have, however, *the same fixed points* (stable configurations).

(3) For a given operator F there is a *unique* parallel mode of operation denoted by $(1, 2, \ldots, n)$. The serial mode of operation (Gauss-Seidel) may similarly be denoted by $((1)(2)\ldots(n))$. There are $n!$ possible serial modes of operation given by the $n!$ permutations of the set $\{1, 2, \ldots, n\}$. Below, we define intermediary modes of operation.

An interesting question (see Chaps. 2, 4, 5) is to try to compare the different modes of operation (for a given operator F) by comparing the *shapes* of the associated iteration graphs.

6. Serial-Parallel Modes of Operation and the Associated Operators

A family of serial-parallel modes of operation is now defined between the purely parallel and the purely serial modes of operation. Basic ideas here go back to the work by Chazan and Miranker, Ref. [21].

Definition. *A serial-parallel mode of operation* or *process* τ for a given automata network is given by an ordered partition $\tau = (\tau_1, \tau_2, \ldots, \tau_s)$ of the set $\{1, 2, \ldots, n\}$.

Example. $n=6$ $\tau = ((4, 1, 6)(2, 3)(5))$.

If two serial-parallel modes of operation τ and γ are given then one says that γ is *more sequential than* τ (written $\tau \alpha \gamma$) if γ can be obtained from τ by *decomposition.*

Example. For $\gamma = ((6, 4)(1)(2, 3)(5))$ clearly $\tau \alpha \gamma$.

The relation α is a *partial ordering* on the set of serial-parallel modes of operation. As an example, let $n=3$. Then the processes $((1, 2)(3))$ and $((2, 3)(1))$ cannot be ordered. The processes $((1, 2)(3))$ and $((1)(2, 3))$ do, however, have a common largest serial process $((1)(2)(3))$ (Gauss-Seidel).

For each serial-parallel process τ we will associate an iteration which is the method of successive approximations for an operator F_τ.

The following example is given to elucidate this:

For $n=6$ and $\tau = ((4, 1, 6)(2, 3)(5))$ the relation $y = F_\tau(x)$ is defined by

$$\begin{array}{l}
y_4 = f_4(x_1, x_2, x_3, x_4, x_5, x_6) \\
y_1 = f_1(x_1, x_2, x_3, x_4, x_5, x_6) \\
y_6 = f_6(x_1, x_2, x_3, x_4, x_5, x_6) \\
\hline
y_2 = f_2(y_1, x_2, x_3, y_4, x_5, y_6) \\
y_3 = f_3(y_1, x_2, x_3, y_4, x_5, y_6) \\
\hline
y_5 = f_5(y_1, y_2, y_3, y_4, x_5, y_6)
\end{array}$$

In the general case, with $\tau=(\tau_1, \tau_2, \ldots, \tau_s)$ the relation $y=F_\tau(x)$ is defined, using an obvious notation, by:

$$\begin{aligned}
y_{\tau_1} &= f_{\tau_1}(x_{\tau_1}, \ldots, x_{\tau_s})\\
y_{\tau_2} &= f_{\tau_2}(y_{\tau_1}, x_{\tau_2}, \ldots, x_{\tau_s})\\
&\vdots\\
y_{\tau_s} &= f_{\tau_s}(y_{\tau_1}, \ldots, y_{\tau_{s-1}}, x_{\tau_s}).
\end{aligned}$$

Definition. F_τ is the operator associated with the operating mode τ.

Remarks. (1) The operator F (associated with the purely parallel mode of operation) is also defined, using $a=(1, 2, \ldots, n)$, by $F_a=F$. Moreover, if $\omega=((1)(2)\ldots(n))$, then the Gauss-Seidel operator G of F is defined by $F_\omega=G$. Clearly we have: $a \propto \omega$.

(2) From the construction it is obvious that *all the operators* F_τ *have the same set of fixed points* (stable configurations). This is furthermore the set (possibly empty) of fixed points of F.

(3) If an initial configuration x^0 is given as well as a serial-parallel mode of operation τ, then to *run the automata network according to the operation mode* τ simply means, by definition, to start with x^0 and to form the successive approximations

$$x^{r+1}=F_\tau(x^r) \qquad (r=0, 1, 2, \ldots).$$

Example. $n=4$ and $\tau=((3, 1)(2)(4))$.

One obtains

$$\begin{array}{l}
x_3^{r+1}=f_3(x_1^r, x_2^r, x_3^r, x_4^r)\\
x_1^{r+1}=f_1(x_1^r, x_2^r, x_3^r, x_4^r)\\
\hline
x_2^{r+1}=f_2(x_1^{r+1}, x_2^r, x_3^{r+1}, x_4^r)\\
\hline
x_4^{r+1}=f_4(x_1^{r+1}, x_2^{r+1}, x_3^{r+1}, x_4^r).
\end{array}$$

Therefore, in the general case, in order to pass from x^r to x^{r+1} according to the serial-parallel mode of operation $\tau=(\tau_1, \tau_2, \ldots, \tau_s)$ one proceeds as follows:

– *parallel* operation of the cells indexed by τ_1

– *then* a *parallel* operation of the cells indexed by τ_2, taking into account the results of the operations made by τ_1

– etc.

This is nothing but the *method of Gauss-Seidel operating on blocks of* F.

It is now easy to show that between two serial-parallel processes τ and γ, the statement that "γ is more sequential than τ" corresponds well to the actual algorithmic modes of operation.

The following theorem is of fundamental importance for the comparison of two serial-parallel modes of operation (regarding their iteration graphs):

Theorem 3. *Let τ and γ be two serial-parallel processes satisfying $\tau \alpha \gamma$. Then F_γ is a blockwise Gauss-Seidel operator associated with F_τ.*

In order to prove this, it suffices to show that for

$$\tau=(\tau_1, \tau_2) \quad \text{and} \quad \gamma=(\tau_1, \tau_2', \tau_2'') \quad \text{with} \quad \begin{cases} \tau_2' \cup \tau_2''=\tau_2 \\ \tau_2' \cap \tau_2''=\emptyset \end{cases}$$

the relation $y=F_\tau(x)$ can be written

$$\begin{cases} y_{\tau_1}=f_{\tau_1}(x_{\tau_1}, x_{\tau_2'}, x_{\tau_2''}) \\ y_{\tau_2'}=f_{\tau_2'}(y_{\tau_1}\;, x_{\tau_2'}, x_{\tau_2''}) \\ y_{\tau_2''}=f_{\tau_2''}(y_{\tau_1}\;, x_{\tau_2'}, x_{\tau_2''}). \end{cases}$$

Then the relation $z=F_\gamma(x)$ can be written

$$\begin{cases} z_{\tau_1}=f_{\tau_1}(x_{\tau_1}, x_{\tau_2'}, x_{\tau_2''}) \\ z_{\tau_2'}=f_{\tau_2'}(z_{\tau_1}\;, x_{\tau_2'}, x_{\tau_2''}) \\ z_{\tau_2''}=f_{\tau_2''}(z_{\tau_1}, z_{\tau_2'}\;, x_{\tau_2''}) \end{cases}$$

from which the result is obtained. □

7. Examples

(1) Let us reconsider the example on p. 10 :

$$X=\{0,1\}^3$$
$$f_1(x)=x_2\,\bar{x}_3$$
$$f_2(x)=x_1\,\bar{x}_3$$
$$f_3(x)=\bar{x}_1\,\bar{x}_2.$$

On p. 92 *all* the iteration graphs for the serial-parallel processes for F are displayed.

(2) $X=\{0,1\}^4=\{a, b, \ldots, p\}$

$$f_1(x)=x_2$$
$$f_2(x)=x_1+\bar{x}_3$$
$$f_3(x)=x_1+x_2$$
$$f_4(x)=1.$$

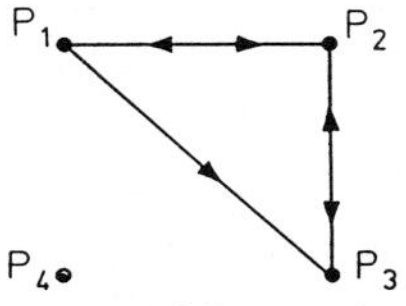

P1 P2 P3 P4

Connectivity graph

Let us also consider, apart from the purely parallel mode of operation $(1,2,3,4)$, the following modes of operation:

Serial ((1) (2) (3) (4))

$$g_1(x)=x_2$$
$$g_2(x)=x_2+\bar{x}_3$$
$$g_3(x)=x_2+x_2+\bar{x}_3=x_2+\bar{x}_3$$
$$g_4(x)=1$$

and the operating mode ((1, 2) (3, 4)) corresponding to the operator H defined by

$$h_1(x)=x_2$$
$$h_2(x)=x_1+\bar{x}_3$$
$$h_3(x)=x_2+x_1+\bar{x}_3$$
$$h_4(x)=1.$$

The tables for these operators and the various iteration graphs are now shown below

	x	$F(x)$		$G(x)$		$H(x)$	
a	0 0 0 0	0 1 0 1	*f*	0 1 1 1	*h*	0 1 1 1	*h*
b	0 0 0 1	0 1 0 1	*f*	0 1 1 1	*h*	0 1 1 1	*h*
c	0 0 1 0	0 0 0 1	*b*	0 0 0 1	*b*	0 0 0 1	*b*
d	0 0 1 1	0 0 0 1	*b*	0 0 0 1	*b*	0 0 0 1	*b*
e	0 1 0 0	1 1 1 1	*p*	1 1 1 1	*p*	1 1 1 1	*p*
f	0 1 0 1	1 1 1 1	*p*	1 1 1 1	*p*	1 1 1 1	*p*
g	0 1 1 0	1 0 1 1	*l*	1 1 0 1	*n*	1 0 1 1	*l*
h	0 1 1 1	1 0 1 1	*l*	1 1 0 1	*n*	1 0 1 1	*l*
i	1 0 0 0	0 1 1 1	*h*	0 1 1 1	*h*	0 1 1 1	*h*
j	1 0 0 1	0 1 1 1	*h*	0 1 1 1	*h*	0 1 1 1	*h*
k	1 0 1 0	0 1 1 1	*h*	0 0 0 1	*b*	0 1 1 1	*h*
l	1 0 1 1	0 1 1 1	*h*	0 0 0 1	*b*	0 1 1 1	*h*
m	1 1 0 0	1 1 1 1	*p*	1 1 1 1	*p*	1 1 1 1	*p*
n	1 1 0 1	1 1 1 1	*p*	1 1 1 1	*p*	1 1 1 1	*p*
o	1 1 1 0	1 1 1 1	*p*	1 1 1 1	*p*	1 1 1 1	*p*
p	1 1 1 1	1 1 1 1	*p*	1 1 1 1	*p*	1 1 1 1	*p*

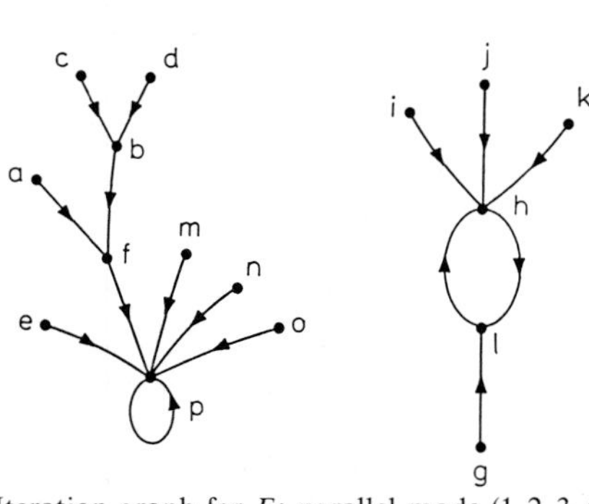

Iteration graph for F; parallel mode (1, 2, 3, 4)

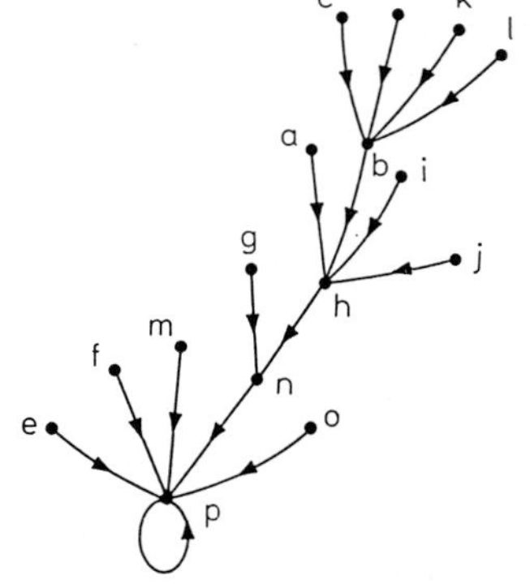

Iteration graph for G; serial mode ((1), (2) (3) (4)) (this graph is simple)

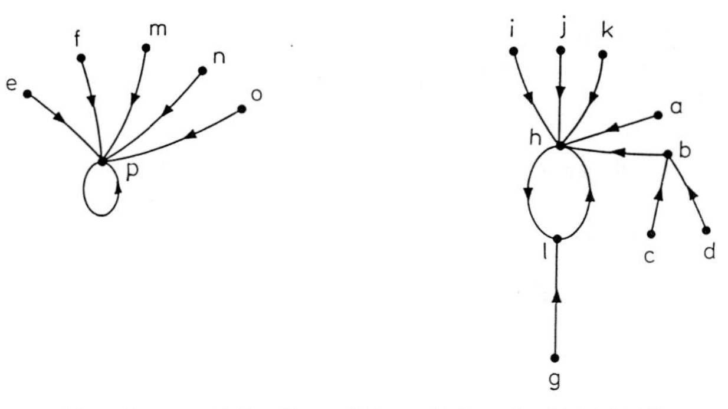

Iteration graph for H; serial-parallel mode ((1, 2) (3, 4))

There are some visual analogies between the various modes of operation.

Clearly, when the number of points gets large, the graphs get very complex.

(3) 'Majority' iterations (see [87]–[96]).

One is given a connectivity graph, where each automaton may take on the two possible states 0 and 1. At each node the transition function is the following (majority function):

The cell takes on a state depending on the majority "vote" of the cells to which it is connected. In the case of a tie vote it keeps its previous state. Here are some examples:

a) *Parallel iteration*

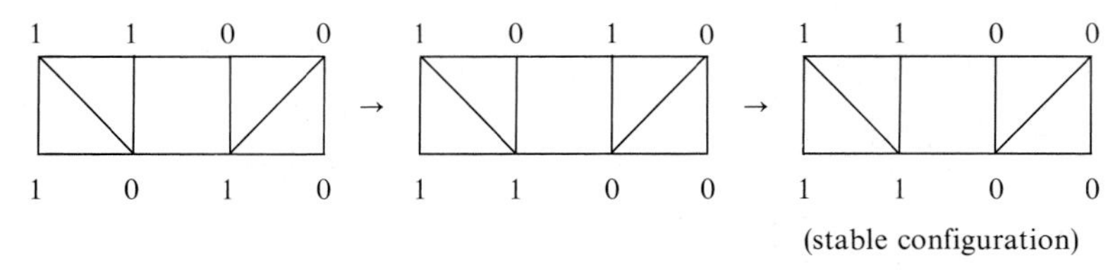

b) *Serial iteration starting from the same initial configuration:*

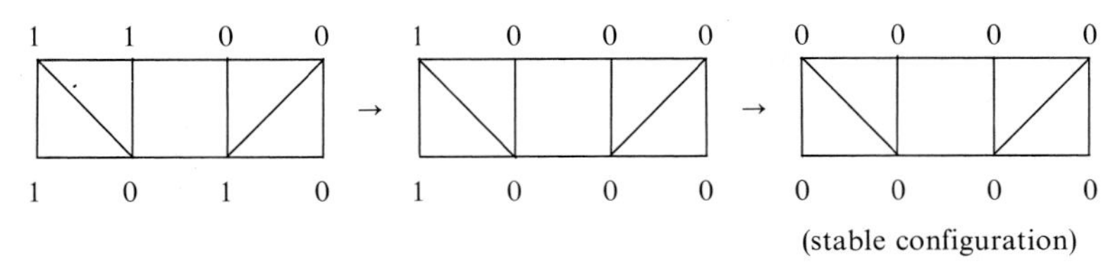

corresponding to the following numbering of the cells:

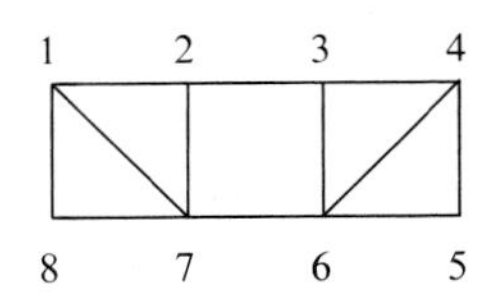

c) For the following numbering:

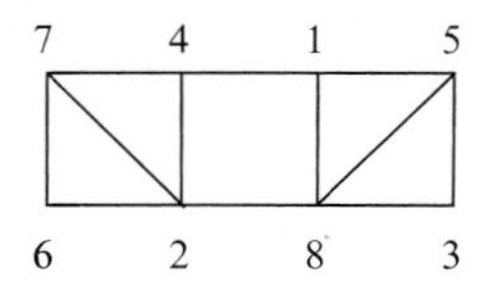

we obtain (serially) with the same initial configuration:

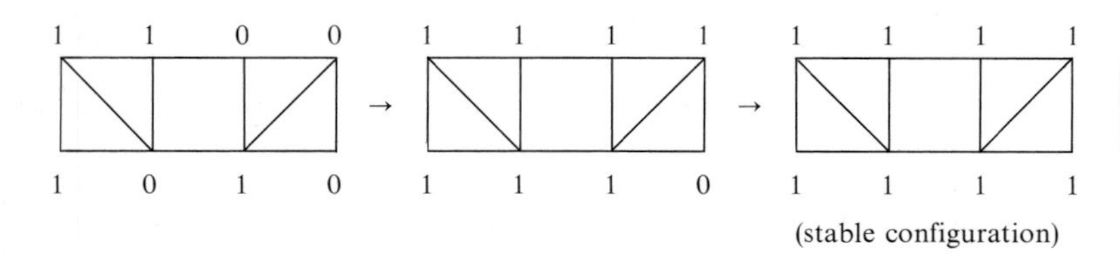

Stated differently, in this example, starting with the same initial configuration, and with the same majority transition function, one obtains, depending on the selected mode of operation, a stable distribution of 0's *and* 1's *in the first case, unanimously* 0 *in the second case and unanimously* 1 *in the third case (and only in* 2 *steps in each case).*

Another example (still with the majority transition function);

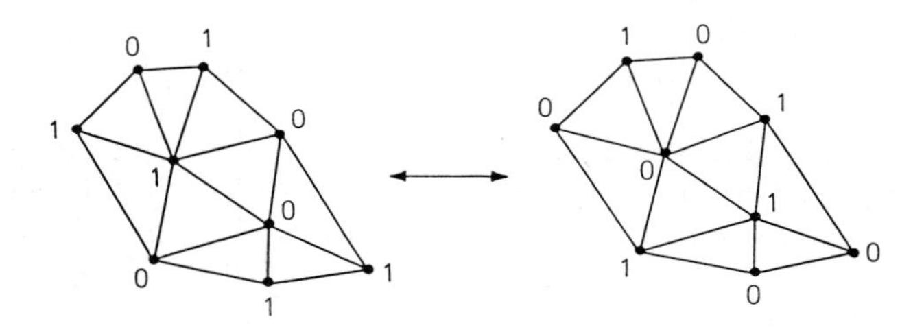

(cycle of length 2 for the parallel mode of operation).

If we use the following numbering:

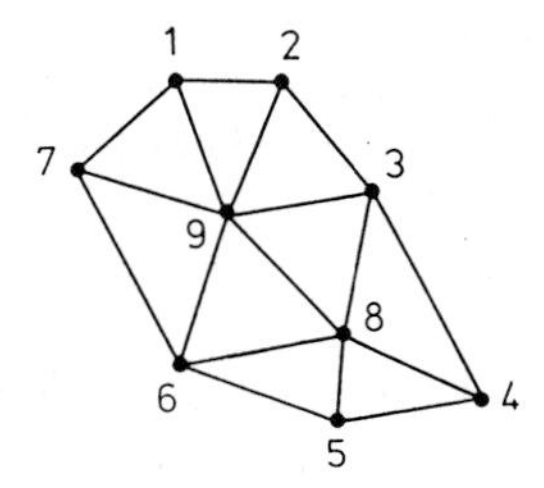

then, starting from the same initial configuration, we obtain a convergence in 2 steps for the corresponding serial iteration:

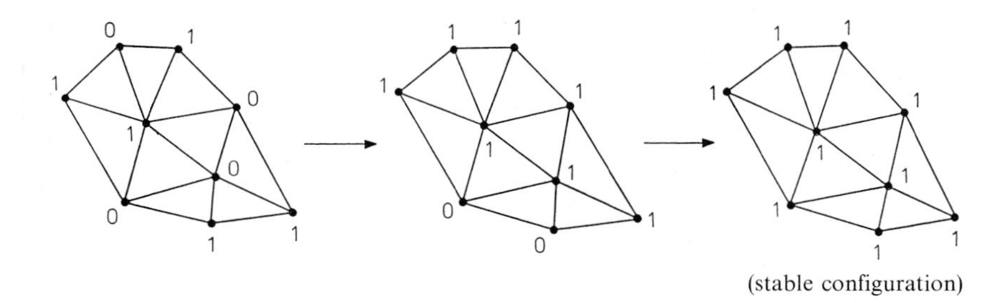

(stable configuration)

Last example (still for the majority transition function)

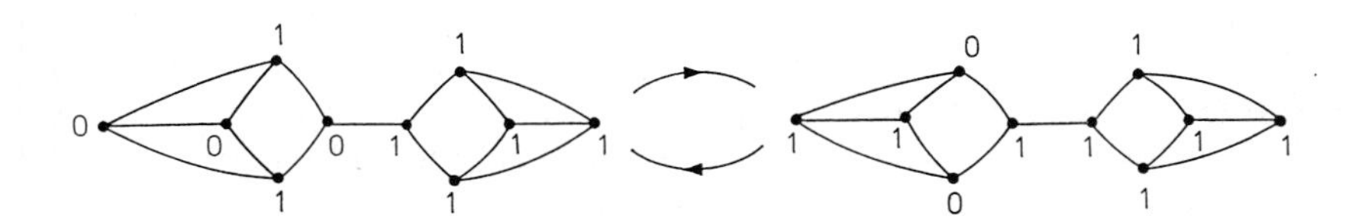

in parallel: a cycle of length 2.

With the same initial configuration, the serial iteration (with some arbitrary numbering of the vertices) always leads to a stable configuration.

Remarks. (1) The following result was obtained by E. Goles [87]–[96]:

Using the majority transition function, having an arbitrary (symmetric*) connectivity graph and an arbitrary initial configuration, the *parallel* operating mode leads to either a fixed point or a cycle of length 2 (*but never to a cycle of length greater than* 2).

A *serial* mode operation *always leads to a fixed point.*

These results hold true for majority functions, of the type defined above. They are also valid for more general transition functions (*threshold* functions, see [87]–[96]).

(2) In the common examples (see Introduction and preceding examples) each transition function f_i only depends on some x_j, or stated differently, each vertex of the connectivity graph for the network has but a few neighbors, which again means that the incidence matrix of F is very *sparse* (see also remark 3 of p. 9). In each line there are only a few non-zero elements.

Here is a typical example:

---

* That is to say that if P_i depends on P_j then P_j must depend on P_i (the connectivity graph is not oriented).

In a square grid of size $(n+2)\times(n+2)$ (where the cells at the edges have a fixed state), the local transition function for a cell (of the $n\times n$ interior cells) only depends on the cell itself and on its four neighboring cells as shown in the following diagram:

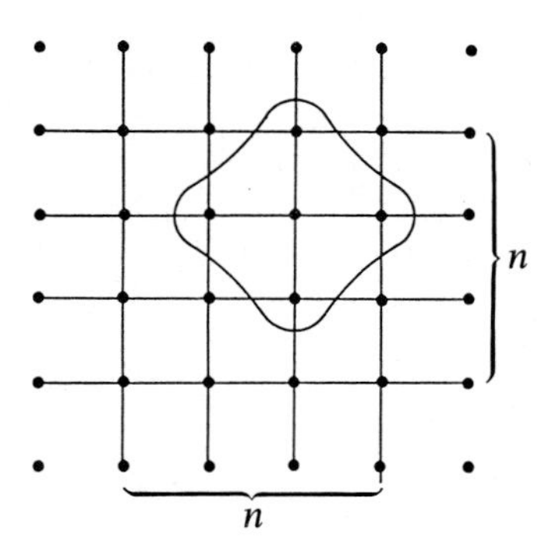

(connectivity graph of the network).

The incidence matrix for the corresponding operator F now has the following form:

$$B(F)=\left[\begin{array}{cccc|cccc|cccc|cccc}
1&1&&&1&&&&&&&&&&&\\
1&1&1&&&1&&&&&&&&&&\\
&1&1&1&&&1&&&&&&&&&\\
&&1&1&&&&1&&&&&&&&\\
\hline
1&&&&1&1&&&1&&&&&&&\\
&1&&&1&1&1&&&1&&&&&&\\
&&1&&&1&1&1&&&1&&&&&\\
&&&1&&&1&1&&&&1&&&&\\
\hline
&&&&1&&&&1&1&&&1&&&\\
&&&&&1&&&1&1&1&&&1&&\\
&&&&&&1&&&1&1&1&&&1&\\
&&&&&&&1&&&1&1&&&&1\\
\hline
&&&&&&&&1&&&&1&1&&\\
&&&&&&&&&1&&&1&1&1&\\
&&&&&&&&&&1&&&1&1&1\\
&&&&&&&&&&&1&&&1&1
\end{array}\right]$$

(elements not shown are zero)

This common example is analogous to the example in numerical analysis when one uses the standard discretization procedures for a problem of the type $\frac{\partial^2 u}{\partial x^2}+\frac{\partial^2 u}{\partial y^2}=f$. The matrix obtained here has the same non-zero elements, and the modes of operation (parallel: Jacobi iteration, serial: Gauss-Seidel) are exactly the same. The essential difference is that within the framework of numerical analysis, the transition function at each node is a map from $\mathbb{R}^5$ to $\mathbb{R}$, whereas within the framework of discrete iterations it is a map of $\{0, 1\}^5$ into $\{0, 1\}$.

It should, however, be noted that all practical calculations are normally done by replacing $\mathbb{R}$ *by a finite set (the set of machine numbers). An execution of a numerical iteration is therefore, in fact, the same as the operation of an automata network.*

It is this analogy that forms the basis for the metric studies of discrete iterations presented in the following chapters.

2. A Metric Tool

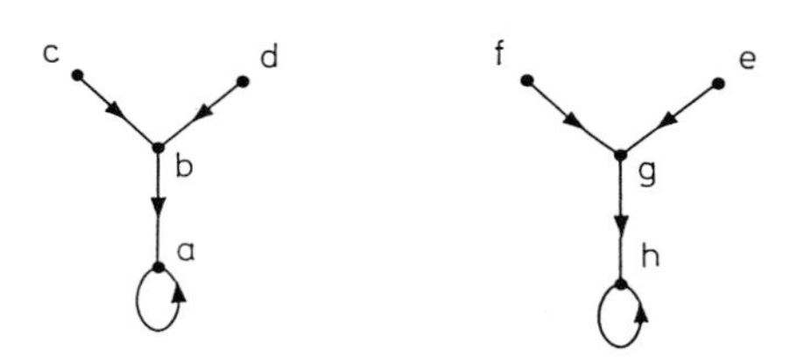

In this chapter we develop a *metric tool* that will enable us to study the behaviour of a discrete iteration in the context of the previous chapter. This tool will be used to define a *distance between two configurations* in order that the convergence of an iteration towards a limit or towards a cycle may be discussed.

The metric tool that is introduced will allow us to systematically transfer the basic results for iterative methods in numerical analysis in the setting of normed spaces into our discrete context throughout the following chapters. It is furthermore this metric that unifies the whole study.

Let us recall the context. $X = \prod_{i=1}^{n} X_i$ is the cartesian product of n finite sets X_i and F is a map of X into X. Starting with $x^0 \in X$, the recurrence relation

$$x^{r+1} = F(x^r) \qquad (r = 0, 1, 2, \ldots)$$

defines a sequence of iterates of F.

In the context of an iteration on a finite set X, the convergence of such a sequence is nothing but (see Chap. 1) a *stationary convergence*, that is the limit is reached in a *finite* number of steps.

The first tool for dealing with this convergence that comes to mind is the *discrete metric* δ on X defined by

$$\delta(x, y) = 1 \quad \text{if} \quad x \neq y; \qquad \delta(x, y) = 0 \quad \text{if} \quad x = y.$$

The convergent sequences in this metric, as well as the Cauchy sequences, coincide with the stationary sequences. (X is complete when equipped with δ.)

It turns out, however, that this metric is much too "coarse" for all purposes in this text. The following discussion illustrates this point:

A sufficient condition for the iteration graph of F to be simple (that is to say that it only contains one basin having a fixed point, see also Chap. 1, p. 6) is clearly that F is *contracting* in the discrete metric δ.

It is easy to show, however, that the only contracting maps F with respect to the metric δ are the *constant* maps, that is to say maps having iteration graphs of the following kind:

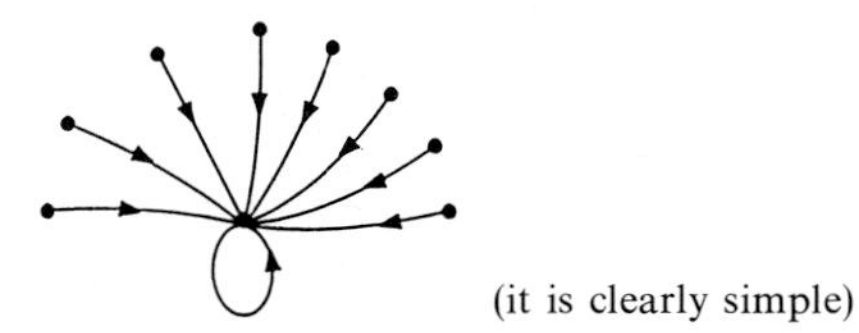

(it is clearly simple)

The above sufficient condition for the iteration graph to be simple is therefore truly a very restrictive condition. The tool that we now introduce is much more flexible.

1. The Boolean vector Distance d

Let $X = \prod_{i=1}^{n} X_i$ be a product of n finite sets X_i. Let δ_i denote the discrete distance on X_i. For $x=(x_1, \ldots, x_n)$ and $y=(y_1, \ldots, y_n)$ in X, we define the *boolean vector*

$$d(x, y) = \begin{bmatrix} \delta_1(x_1, y_1) \\ \vdots \\ \delta_n(x_n, y_n) \end{bmatrix} \in \{0, 1\}^n.$$

d satisfies the following axioms:

$$d(x, y) = 0 \Rightarrow x = y$$

$$d(x, y) = d(y, x) \quad \forall\, x, y \in X$$

$$d(x, z) \leq d(x, y) + d(y, z) \quad \forall\, x, y, z \in X.$$

Beware! In the last line, the $+$ represents the *boolean sum* of vectors in $\{0, 1\}^n$ $(1+1=1)$, while *the inequality that appears is the componentwise inequality* for the order relation $0 \leq 0 \leq 1 \leq 1$ in the boolean algebra $\{0, 1\}$.

Example. $X=\{0,1\}^4$

For $x=(0\ 0\ 1\ 1)$ and $y=(0\ 1\ 0\ 1)$ one obtains

$$d(x,y)=\begin{bmatrix}0\\1\\1\\0\end{bmatrix}.$$

Moreover, for $z=(0\ 0\ 0\ 0)$ one has

$$d(x,z)=\begin{bmatrix}0\\0\\1\\1\end{bmatrix},\qquad d(y,z)=\begin{bmatrix}0\\1\\0\\1\end{bmatrix}$$

and one verifies that

$$d(x,z)=\begin{bmatrix}0\\0\\1\\1\end{bmatrix}\leq d(x,y)+d(y,z)=\begin{bmatrix}0\\1\\1\\0\end{bmatrix}+\begin{bmatrix}0\\1\\0\\1\end{bmatrix}=\begin{bmatrix}0\\1\\1\\1\end{bmatrix}.$$

Even though $d(x,y)$ is a boolean vector (and intentionally not a vector ≥ 0 in $\mathbb{R}^n$), *d constitutes a metric tool* on X *which we call the boolean vector distance on X*. It is equivalent to the discrete metric on X. Indeed for both tools, the Cauchy sequences and the convergent sequences are the stationary sequences (which gives the completion of X by d!).

We are interested in using the metric d, because with the aid of this metric, we are able to carry over to our metric space (X,d) the notions of analysis (contraction, local convergence, attractive fixed points and cycles, Newton's method, etc.) *and thus obtain "non-trivial" results on the iterative behaviour of the map F in the context of an iteration on a product of finite sets.*

2. Some Basic Inequalities

We will start off by proving some elementary results of fundamental importance for this study.

Theorem 1. *For all x and y in X the following inequality is valid componentwise in $\{0,1\}^n$:*

$$d(F(x),F(y))\leq B(F)\,d(x,y)$$

where $B(F)$ is the incidence matrix for F.

Indeed (boolean operations)

$$\begin{aligned}\delta_i(f_i(x_1,\dots,x_n), f_i(y_1,\dots,y_n)) \leq\ & \delta_i(f_i(x_1,\dots,x_n), f_i(y_1,x_2,\dots,x_n))\\ &+\delta_i(f_i(y_1,x_2,\dots,x_n), f_i(y_1,y_2,x_3,\dots,x_n))\\ &+\dots\\ &+\delta_i(f_i(y_1,\dots,y_{n-1},x_n), f_i(y_1,\dots,y_{n-1},y_n))\\ \leq b_{i1}\,\delta_1(x_1,y_1)+b_{i2}\,\delta_2(x_2,y_2)+&\dots+b_{in}\,\delta_n(x_n,y_n)\end{aligned}$$

where b_{ij} denotes the element in the position (i,j) in the incidence matrix $B(F)$. (The last inequality follows from the definition of $B(F)$.)

Since this is valid for all i, $1 \leq i \leq n$, the result is proven. □

(The interest in the metric tool d on X now arises from the fact that the above inequality is valid for any F mapping X into itself.)

Example (see Chap. 1, p. 8)

$$n=3 \quad X_i=\{0,1\} \quad (i=1,2,3) \quad X=\{0,1\}^3$$

and F is given by the table

x	$F(x)$
0 0 0	0 1 0
0 0 1	1 0 0
0 1 0	0 1 0
0 1 1	1 0 1
1 0 0	1 1 0
1 0 1	1 1 0
1 1 0	0 1 0
1 1 1	1 1 1

with $B(F)=\begin{bmatrix}1&1&1\\1&0&1\\0&1&1\end{bmatrix}$.

Let $x=(0\ 0\ 1)$ and $y=(0\ 1\ 1)$ then

$$F(x)=(1\ 0\ 0) \quad \text{and} \quad F(y)=(1\ 0\ 1).$$

The inequality of Theorem 1 is now written as

$$d(F(x),F(y))=\begin{bmatrix}0\\0\\1\end{bmatrix} \leq \begin{bmatrix}1&1&1\\1&0&1\\0&1&1\end{bmatrix}\begin{bmatrix}0\\1\\0\end{bmatrix}=B(F)\,d(x,y)=\begin{bmatrix}1\\0\\1\end{bmatrix}.$$

Theorem 2. *In order that a boolean $n\times n$ matrix M satisfies*

$$\forall\, x,y\in X \qquad d(F(x),F(y)) \leq M\,d(x,y)$$

it is necessary and sufficient that

$$B(F) \leq M$$

(where the inequality is elementwise between $B(F)$ and M using the order relation $0 \le 0 \le 1 \le 1$ in $\{0, 1\}$).

The sufficient condition is obvious:

$$d(F(x), F(y)) \le B(F)\, d(x, y) \le M\, d(x, y).$$

As for the necessary condition we suppose that a boolean $n \times n$ matrix $M = (m_{ij})$ satisfies the inequality

$$\forall x, y \in X \qquad d(F(x), F(y)) \le M\, d(x, y) \tag{1}$$

while also having at least one element m_{ij} for which $m_{ij} < b_{ij}$, where b_{ij} is the corresponding element of $B(F)$. This means that $b_{ij} = 1$ and $m_{ij} = 0$.

Since $b_{ij} = 1$ it follows that f_i depends on x_j. Otherwise stated, using the definition of $B(F)$, we have that there exists $x = (x_1, \ldots, x_j, \ldots, x_n)$ and $x' = (x_1, \ldots, y_j, \ldots, x_n)$ with $y_j \neq x_j$ in X_j such that

$$\delta_i(f_i(x), f_i(x')) = 1.$$

So, using (1), we may majorize this quantity by

$$\sum_{s=1}^{j-1} m_{is} \underbrace{\delta_s(x_s, x_s)}_{0} + \underbrace{m_{ij}}_{0} \underbrace{\delta_j(x_j, y_j)}_{1} + \sum_{s=j+1}^{n} m_{is} \underbrace{\delta_s(x_s, x_s)}_{0} = 0$$

which leads to a contradiction, that is, it follows that $m_{ij} \ge b_{ij}$ $(i, j = 1, 2, \ldots, n)$. □

Theorem 3. *Let E and F be maps of X into itself. Then it follows that*

$$B(E \circ F) \le B(E)\, B(F) \qquad \textit{(boolean matrix product)}.$$

The proof is elementary, starting from the inequalities

$$\forall x, y \in X \qquad d(E \circ F(x), E \circ F(y)) \le B(E)\, d(F(x), F(y)) \le B(E)\, B(F)\, d(x, y)$$

and then using the preceding theorem. □

3. First Applications

We now present some immediate applications of the preceding three theorems:

Theorem 4. *Let ξ be a fixed point for F and let $x^r = F^r(x^0)$ be the r-th iterate starting from x^0. Then we have*

$$d(x^r, \xi) \le [B(F)]^r\, d(x^0, \xi).$$

Clearly, for each x and y in X one has

$$d(F^r(x), F^r(y)) \leq [B(F)]^r \, d(x, y)$$

after repeatedly applying Theorems 1 and 3. With $y = \xi$, the fixed point of F, and with $x = x^0$, the result is obtained. □

Corollary. *If* $[B(F)]^r$ *(boolean power) has columns indexed* $j_1, j_2, \ldots, j_s$ *equal to zero, then for all* x^0 *that only differs from a fixed point* ξ *of* F *in one or more of the components indexed by* $j_1, j_2, \ldots, j_s$ *it follows that* $x^r = \xi$.

This result follows immediately when the preceding inequality is applied

$$0 \leq d(x^r, \xi) \leq [B(F)]^r \, d(x^0, \xi) = 0$$

from which $x^r = \xi$. □

Example. $n = 3$.

Let us take

$$B(F) = \begin{bmatrix} 0 & 1 & 1 \\ 0 & 0 & 1 \\ 0 & 0 & 1 \end{bmatrix}$$

with, for example, $X = \{0, 1\}^3$ and F defined by

$$f_1(x) = x_2 + \bar{x}_3$$
$$f_2(x) = \bar{x}_3$$
$$f_3(x) = x_3.$$

The table for F and its iteration graph are then given by:

	x			$F(x)$			
a	0	0	0	1	1	0	g
b	0	0	1	0	0	1	b
c	0	1	0	1	1	0	g
d	0	1	1	1	0	1	f
e	1	0	0	1	1	0	g
f	1	0	1	0	0	1	b
g	1	1	0	1	1	0	g
h	1	1	1	1	0	1	f

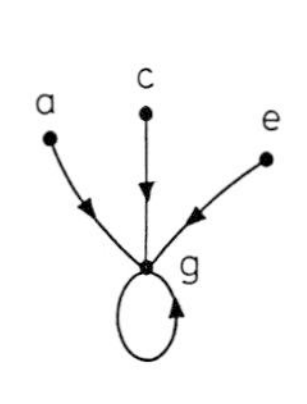

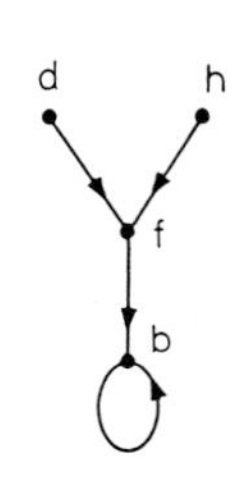

Let us consider the fixed point $b = (0 \ 0 \ 1)$ of F.

a) $r = 1$. $B(F)$ has the first column equal to zero. $f = (1 \ 0 \ 1)$ only differs from b in the first component. One verifies that for $x^0 = f$, $x^1 = b$.

b) $r=2$. $[B(F)]^2=\begin{bmatrix}0&0&1\\0&0&1\\0&0&1\end{bmatrix}$ has the first two columns equal to zero, h differs from b in the first two components. It is easily verified that $F^2(h)=b$.

d only differs from b in the second component. One easily verifies that $F^2(d)=b$.

Remark. In the context of the preceding theorem, suppose that F has a cycle of length p, i.e. $\xi_1, \xi_2, \ldots, \xi_p$. Then all ξ_i are fixed points of F^p, and in applying the preceding corollary to F^p, one may establish a sufficient condition for the iteration $x^{r+1}=F(x^r)$ to arrive at an element of the cycle in $r \cdot p$ steps.

We now consider the Gauss-Seidel operator associated with F (see Chap. 1, p. 11) and we establish certain basic inequalities.

First of all remember (see Chap. 1, p. 12) that G is written as

$$G=F_n \circ \ldots \circ F_2 \circ F_1$$

with

$$F_i(x)=\begin{bmatrix}x_1\\ \vdots \\ f_i(x_1 \ldots x_n)\\ \vdots \\ x_n\end{bmatrix}.$$

Denoting the elements of $B(F)$ by b_{ij} one has

$$B(F_i)=\begin{bmatrix}1 & & & & & \\ & \ddots & & & & \\ & & 1 & & & \\ b_{i1} & \ldots & b_{ii} & \ldots & b_{in} \\ & & & 1 & & \\ & & & & \ddots & \\ & & & & & 1\end{bmatrix} \qquad (i=1, 2, \ldots, n)$$

(elements not shown are zero).

Theorem 5.

$$B(G) \leq B(F_n) \ldots B(F_1).$$

The result follows immediately by applying Theorem 3 above repeatedly. □

Theorem 6. *Let L be the strictly lower part of $B(F)$ and U the upper triangular part. Then the inequality*

$$\forall\, x, y \in X \qquad d(G(x), G(y)) \leq L\, d(G(x), G(y)) + U\, d(x, y)$$

holds as well as the inequality

$$B(G) \le [I + L + \ldots + L^{n-1}]\, U.$$

From Theorem 1 it follows that

$$\delta_i(f_i(x), f_i(y)) \le \sum_{j=1}^{n} b_{ij}\, \delta_j(x_j, y_j).$$

Then from the definition of G we get

$$\forall\, x, y \in X \qquad \delta_i(g_i(x), g_i(y)) \le \sum_{j=1}^{i-1} b_{ij}\, \delta_j(g_j(x), g_j(y)) + \sum_{j=i}^{n} b_{ij}\, \delta_j(x_j, y_j)$$
$$(i = 1, 2, \ldots, n).$$

By regrouping, the following inequality is obtained:

$$\forall\, x, y \in X \qquad d(G(x), G(y)) \le L\, d(G(x), G(y)) + U\, d(x, y).$$

Now let $\alpha = d(x, y)$ and $\beta = d(G(x), G(y))$ for simplicity. Then

$$\beta \le L\,\beta + U\,\alpha$$

and by iterating we get

$$\beta \le L(L\,\beta + U\,\alpha) + U\,\alpha = [LU + U]\,\alpha + L^2\,\beta$$

and in general for any r

$$\beta \le [I + L + \ldots + L^{r-1}]\, U\,\alpha + L^r\,\beta \qquad \text{(boolean powers)}.$$

Since L is *strictly* lower triangular it follows for all $r \ge n$ that

$$L^r = 0 \qquad \text{(boolean powers)},$$

from which the inequality

$$\beta \le [I + L + \ldots + L^{n-1}]\, U\,\alpha$$

is obtained.

Finally, one has

$$\forall\, x, y \in X \qquad d(G(x), G(y)) \le [I + L + \ldots + L^{n-1}]\, U\, d(x, y)$$

from which it follows that (see Theorem 2)

$$B(G) \le [I + L + \ldots + L^{n-1}]\, U. \qquad \square$$

Remark. In comparing the two majorizing expressions for $B(G)$ established by Theorems 5 and 6, one may prove that

$$B(F_n) \ldots B(F_1) = [I + L + \ldots + L^{n-1}]\, U.$$

This majorizing expression for $B(G)$ is used heavily in Chap. 4.

Example. Remembering the example on p. 8 one obtains

$$B(F)=\begin{bmatrix}1&1&1\\1&0&1\\0&1&1\end{bmatrix} \quad \text{from which} \quad L=\begin{bmatrix}0&0&0\\1&0&0\\0&1&0\end{bmatrix} \quad \text{and} \quad U=\begin{bmatrix}1&1&1\\0&0&1\\0&0&1\end{bmatrix}.$$

With

$$B(F_1)=\begin{bmatrix}1&1&1\\0&1&0\\0&0&1\end{bmatrix}, \quad B(F_2)=\begin{bmatrix}1&0&0\\1&0&1\\0&0&1\end{bmatrix}, \quad B(F_3)=\begin{bmatrix}1&0&0\\0&1&0\\0&1&1\end{bmatrix}$$

one gets

$$B(F_3)\,B(F_2)\,B(F_1)=(I+L+L^2)\,U=\begin{bmatrix}1&1&1\\1&1&1\\1&1&1\end{bmatrix}.$$

In this example, starting with the table for F, one may construct the table for G

x			$F(x)$			$G(x)$		
0	0	0	0	1	0	0	1	0
0	0	1	1	0	0	1	1	1
0	1	0	0	1	0	0	1	0
0	1	1	1	0	1	1	1	1
1	0	0	1	1	0	1	1	0
1	0	1	1	1	0	1	1	1
1	1	0	0	1	0	0	1	0
1	1	1	1	1	1	1	1	1

Moreover, one has (see pp. 8 and 12) that

$$\begin{aligned} f_1(x)&=x_1\bar{x}_2+x_3 & & & g_1(x)&=x_1\bar{x}_2+x_3\\ f_2(x)&=x_1+\bar{x}_3 & &\text{which implies} & g_2(x)&=1\\ f_3(x)&=x_2x_3 & & & g_3(x)&=x_3. \end{aligned}$$

It is easily verified that

$$B(G)=\begin{bmatrix}1&1&1\\0&0&0\\0&0&1\end{bmatrix}\le(I+L+L^2)\,U=\begin{bmatrix}1&1&1\\1&1&1\\1&1&1\end{bmatrix}.$$

The inequality is in this case not informative since the matrix $[I+L+\ldots+L^{n-1}]\,U$ contains only ones.

4. Serial-Parallel Operators. An Outline

Results analogous to the results in Theorems 5 and 6 for G may be established without difficulty for the serial-parallel operators defined starting from F. We only give an outline of the main ideas here in order that the exposition shall not be burdened with too much detail.

If $\tau=(\tau_1, \tau_2, \ldots, \tau_s)$ is a serial-parallel process then the associated operator F_τ may also be factorized (in the same manner that G was factorized into $F_n \circ \ldots \circ F_2 \circ F_1$).

Now, introducing the operators F_{τ_i} defined by (notation as on p. 17), the relation $y=F_{\tau_i}(x)$ being written as

$$\begin{cases} y_{\tau_r}=x_{\tau_r} & \text{for } r \neq i \\ y_{\tau_i}=F_{\tau_i}(x)=F_{\tau_i}(x_{\tau_1}, \ldots, x_{\tau_s}), \end{cases}$$

one may factorize F_τ as follows

$$F_\tau=F_{\tau_s} \circ \ldots \circ F_{\tau_2} \circ F_{\tau_1}$$

from which the inequality

$$B(F_\tau) \leq B(F_{\tau_s}) \ldots B(F_{\tau_2}) B(F_{\tau_1})$$

is obtained.

Remark. $F_\tau=G$ for $\tau_i=(i)$ $i=1,2, \ldots, n$.

Example. Let us reconsider the example on p. 17

$$n=6 \quad \text{and} \quad \tau_1=(4,1,6) \quad \tau_2=(2,3) \quad \tau_3=(5).$$

The operator F_{τ_1} is defined for $y=F_{\tau_1}(x)$ by

$$\begin{aligned} & y_1=f_1(x_1, x_2, x_3, x_4, x_5, x_6) \\ & y_2=x_2 \\ & y_3=x_3 \\ & y_4=f_4(x_1, x_2, x_3, x_4, x_5, x_6) \\ & y_5=x_5 \\ & y_6=f_6(x_1, x_2, x_3, x_4, x_5, x_6). \end{aligned}$$

Then

$$B(F_{\tau_1})=\begin{bmatrix} b_{11} & b_{12} & b_{13} & b_{14} & b_{15} & b_{16} \\ 0 & 1 & 0 & 0 & 0 & 0 \\ 0 & 0 & 1 & 0 & 0 & 0 \\ b_{41} & b_{42} & b_{43} & b_{44} & b_{45} & b_{46} \\ 0 & 0 & 0 & 0 & 1 & 0 \\ b_{61} & b_{62} & b_{63} & b_{64} & b_{65} & b_{66} \end{bmatrix}$$

F_{τ_2}, F_{τ_3} are defined analogously, from which we get $B(F_{\tau_2})$ and $B(F_{\tau_3})$. The associated operator F_τ may therefore be factorized as:

$$F_\tau = F_{\tau_3} \circ F_{\tau_2} \circ F_{\tau_1}$$

from which

$$B(F_\tau) \leq B(F_{\tau_3})\, B(F_{\tau_2})\, B(F_{\tau_1}).$$

It is moreover remarked on p. 18, Chap. 1 that F_τ is nothing but the Gauss-Seidel operator on associated blocks of F.

Instead of decomposing $B(F)$ into strictly lower triangular and upper triangular matrices, we decompose it here into a sum of two matrices with block structure (according to the blocks considered) also called L and U.

Rather than formalizing this we turn to the preceding example

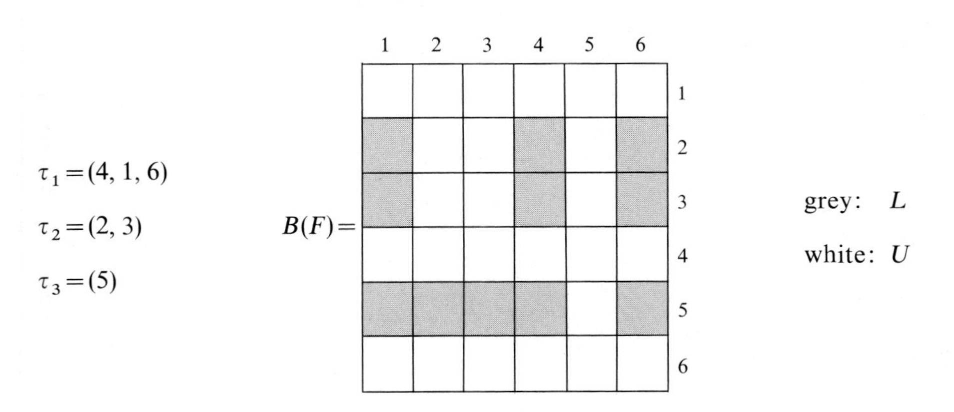

$$B(F) = L + U.$$

One still has (see Theorem 6) in the general case that

$$\forall x, y \in X \qquad d(F_\tau(x), F_\tau(y)) \leq L\, d(F_\tau(x), F_\tau(y)) + U\, d(x, y)$$

and

$$B(F_\tau) \leq [I + L + \ldots + L^{n-1}]\, U = B(F_{\tau_s}) \ldots B(F_{\tau_2})\, B(F_{\tau_1}).$$

Here is a typical numerical example. Let

$$B(F) = \begin{bmatrix} 0 & 0 & 0 & 1 & 1 & 0 \\ 1 & 0 & 1 & 0 & 0 & 1 \\ 0 & 1 & 0 & 1 & 0 & 1 \\ 0 & 0 & 0 & 1 & 0 & 0 \\ 1 & 1 & 0 & 1 & 0 & 1 \\ 0 & 0 & 0 & 1 & 1 & 0 \end{bmatrix}$$

with (for example) $X=\{0,1\}^6$ and (using boolean notation)

$$
\begin{aligned}
f_1(x)&=x_4x_5\\
f_2(x)&=x_1+x_3+x_6\\
f_3(x)&=x_2+x_4+x_6\\
f_4(x)&=x_4\\
f_5(x)&=x_1x_2+x_4+x_6\\
f_6(x)&=x_4x_5.
\end{aligned}
$$

Then, always using $\tau=((4,1,6)\,(2,3)\,(5))$, the operator $F_\tau=H$, is defined by

$$
\begin{aligned}
h_1(x)&=x_4x_5\\
h_2(x)&=x_4x_5+x_3+x_4x_5\\
h_3(x)&=x_2+x_4+x_4x_5\\
h_4(x)&=x_4\\
h_5(x)&=x_4x_5(x_4x_5+x_3+x_4x_5)+x_4+x_4x_5\\
h_6(x)&=x_4x_5
\end{aligned}
$$

which may be somewhat simplified to

$$
\begin{aligned}
h_1(x)&=x_4x_5\\
h_2(x)&=x_3+x_4x_5\\
h_3(x)&=x_2+x_4\\
h_4(x)&=x_4\\
h_5(x)&=x_4\\
h_6(x)&=x_4x_5.
\end{aligned}
$$

In this example one has

$$
B(F_\tau)=\begin{bmatrix}
0&0&0&1&1&0\\
0&0&1&1&1&0\\
0&1&0&1&0&0\\
0&0&0&1&0&0\\
0&0&0&1&0&0\\
0&0&0&1&1&0
\end{bmatrix}
$$

$$
B(F_{\tau_1})=\begin{bmatrix}
0&0&0&1&1&0\\
0&1&0&0&0&0\\
0&0&1&0&0&0\\
0&0&0&1&0&0\\
0&0&0&0&1&0\\
0&0&0&1&1&0
\end{bmatrix}\begin{matrix}*\\ \\ \\ *\\ \\ *\end{matrix}
\qquad
B(F_{\tau_2})=\begin{bmatrix}
1&0&0&0&0&0\\
1&0&1&0&0&1\\
0&1&0&1&0&1\\
0&0&0&1&0&0\\
0&0&0&0&1&0\\
0&0&0&0&0&1
\end{bmatrix}\begin{matrix} \\ *\\ *\\ \\ \\ \end{matrix}
$$

$$B(F_{\tau_3}) = \begin{bmatrix} 1 & 0 & 0 & 0 & 0 & 0 \\ 0 & 1 & 0 & 0 & 0 & 0 \\ 0 & 0 & 1 & 0 & 0 & 0 \\ 0 & 0 & 0 & 1 & 0 & 0 \\ 1 & 1 & 0 & 1 & 0 & 1 \\ 0 & 0 & 0 & 0 & 0 & 1 \end{bmatrix} \begin{matrix} \\ \\ \\ \\ * \\ \\ \end{matrix} .$$

One verifies that

$$B(F_\tau) \leq B(F_{\tau_3})\, B(F_{\tau_2})\, B(F_{\tau_1}) = \begin{bmatrix} 0 & 0 & 0 & 1 & 1 & 0 \\ 0 & 0 & 1 & 1 & 1 & 0 \\ 0 & 1 & 0 & 1 & \textcircled{1} & 0 \\ 0 & 0 & 0 & 1 & 0 & 0 \\ 0 & 0 & \textcircled{1} & 1 & \textcircled{1} & 0 \\ 0 & 0 & 0 & 1 & 1 & 0 \end{bmatrix}.$$

The elements in the last matrix which have grown compared to $B(F_\tau)$ are circled.

Moreover we have

$$L = \begin{bmatrix} 0 & 0 & 0 & 0 & 0 & 0 \\ 1 & 0 & 0 & 0 & 0 & 1 \\ 0 & 0 & 0 & 1 & 0 & 1 \\ 0 & 0 & 0 & 0 & 0 & 0 \\ 1 & 1 & 0 & 1 & 0 & 1 \\ 0 & 0 & 0 & 0 & 0 & 0 \end{bmatrix} \qquad U = \begin{bmatrix} 0 & 0 & 0 & 1 & 1 & 0 \\ 0 & 0 & 1 & 0 & 0 & 0 \\ 0 & 1 & 0 & 0 & 0 & 0 \\ 0 & 0 & 0 & 1 & 0 & 0 \\ 0 & 0 & 0 & 0 & 0 & 0 \\ 0 & 0 & 0 & 1 & 1 & 0 \end{bmatrix}.$$

Then for

$$x = (0\ 0\ 1\ 0\ 1\ 0) \quad \text{we get} \quad H(x) = (0\ 1\ 0\ 0\ 0\ 0)$$

and for

$$y = (0\ 0\ 0\ 0\ 1\ 1) \quad \text{we get} \quad H(y) = (0\ 0\ 0\ 0\ 0\ 0).$$

It is easily verified that

$$d(H(x), H(y)) \leq L\, d(H(x), H(y)) + U\, d(x, y)$$

since

$$\begin{bmatrix} 0 \\ 1 \\ 0 \\ 0 \\ 0 \\ 0 \end{bmatrix} \leq \begin{bmatrix} 0 \\ 0 \\ 0 \\ 0 \\ 1 \\ 0 \end{bmatrix} + \begin{bmatrix} 0 \\ 1 \\ 0 \\ 0 \\ 0 \\ 0 \end{bmatrix} = \begin{bmatrix} 0 \\ 1 \\ 0 \\ 0 \\ 1 \\ 0 \end{bmatrix}.$$

Remarks. (1) In this example, it is easy to see why $B(F_\tau)$ does not coincide with $B(F_{\tau_3})\, B(F_{\tau_2})\, B(F_{\tau_1})$ in general, but only satisfies the inequality $\leq$. During the construction of F_τ certain variables disappear. For example, one would a priori expect x_5 would be present in h_3 (see the first form), but it is eliminated in the final form.

(2) It was already noted in Chap. 1 that $B(F)$ only gives a (partial) information for F, of the type all or nothing. Indeed there are many F having the same $B(F)$. *The preceding inequalities therefore give information valid for all F with the same* $B(F)$. This information is therefore of a global type, sometimes rudimentary, but sometimes good enough.

5. Other Possible Metric Tools

It is possible to "condense" the information given by $B(F)$ further by "condensing" the vector distance used. We will, however, only illustrate this point by an example.

In $X=\{0,1\}^6$, the distance d is indeed defined in the following manner:

$$d(x,y)=\begin{bmatrix}\delta_1(x_1,y_1)\\ \vdots \\ \delta_6(x_6,y_6)\end{bmatrix}.$$

This distance d may now be "condensed" by using, for example, the following distance:

$$\gamma(x,y)=\begin{bmatrix}\mathrm{Max}(\delta_1(x_1,y_1),\delta_2(x_2,y_2),\delta_3(x_3,y_3))\\ \mathrm{Max}(\delta_4(x_4,y_4),\delta_5(x_5,y_5))\\ \delta_6(x_6,y_6)\end{bmatrix}$$

γ is still a boolean vector distance on $\{0,1\}^6$.

For F: $\{0,1\}^6\to\{0,1\}^6$, the basic inequality

$$\forall x,y\in\{0,1\}^6 \qquad d(F(x),F(y))\leq B(F)\,d(x,y)$$

may then be condensed to

$$\forall x,y\in\{0,1\}^6 \qquad \gamma(F(x),F(y))\leq M\,\gamma(x,y)$$

where the matrix M is obtained as follows: $B(F)$ is decomposed into blocks (corresponding to the decomposition into blocks introduced on X).

The element m_{ij} is zero if the corresponding block in $B(F)$ is zero, otherwise it is 1.

If one takes the operator F used in the above example, then

$$B(F)=\left[\begin{array}{ccc|cc|c} 0&0&0&1&1&0\\ 1&0&1&0&0&1\\ 0&1&0&1&0&1\\ \hline 0&0&0&1&0&0\\ 1&1&0&1&0&1\\ \hline 0&0&0&1&1&0 \end{array}\right] \quad \text{from which} \quad M=\begin{bmatrix} 1&1&1\\ 1&1&1\\ 0&1&0 \end{bmatrix}.$$

Take for example

$$x=(0\ 0\ 1\ 0\ 1\ 0) \quad \text{from which} \quad F(x)=(0\ 1\ 0\ 0\ 0\ 0)$$

and

$$y=(0\ 0\ 0\ 0\ 1\ 1) \quad \text{from which} \quad F(y)=(0\ 1\ 1\ 0\ 1\ 0).$$

The inequality $d(F(x), F(y)) \leq B(F)\, d(x, y)$ can be written as

$$\left[\begin{array}{c} 0\\0\\1\\ \hline 0\\1\\ \hline 0 \end{array}\right] \leq \left[\begin{array}{ccc|cc|c} 0&0&0&1&1&0\\ 1&0&1&0&0&1\\ 0&1&0&1&0&1\\ \hline 0&0&0&1&0&0\\ 1&1&0&1&0&1\\ \hline 0&0&0&1&1&0 \end{array}\right] \left[\begin{array}{c} 0\\0\\1\\ \hline 0\\0\\ \hline 1 \end{array}\right] = \left[\begin{array}{c} 0\\1\\1\\ \hline 0\\1\\ \hline 0 \end{array}\right]$$

and then condensed to

$$\gamma(F(x), F(y)) \leq M\, \gamma(x, y)$$

which can be written as

$$\begin{bmatrix} 1\\1\\0 \end{bmatrix} \leq \begin{bmatrix} 1&1&1\\ 1&1&1\\ 0&1&0 \end{bmatrix} \begin{bmatrix} 1\\0\\1 \end{bmatrix} = \begin{bmatrix} 1\\1\\0 \end{bmatrix}.$$

Remarks. (1) The usage of the more compact notation (using γ) results of course in a loss of information (compared to using d).

(2) In the remaining chapters we use the metric d (for reasons of simplicity). All the results obtained using d could easily be transferred if a more compact vector distance was employed.

(3) In the limit, *using the maximum possible condensation, the discrete metric for X is obtained*. This metric is truly the "coarsest" metric (see also the beginning of this chapter, p. 28).

(4) There is therefore the possibility of adjusting the "size" (and, hence the fineness) of the metric used between 1 (the discrete metric) and n (the boolean vector distance d utilized henceforth).

3. The Boolean Perron-Frobenius and Stein-Rosenberg Theorems

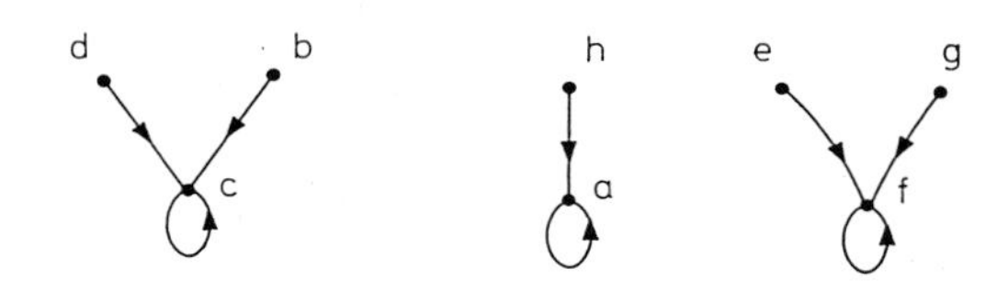

Two theorems are basic to the study of the spectral properties of real square matrices with elements ≥ 0. These are the Perron-Frobenius [24], [25], [38], [44] and the Stein-Rosenberg [17] to [19], [25], [38], [44] theorems. These two theorems form the basic tools for studying the convergence of iterative processes in $\mathbb{R}^n$, linear [25], [44] or non-linear [23], [31] to [33], [37] to [40].

In order to study the behaviour of discrete iterations we pose the question of the feasibility of proving analogous "boolean" versions of these theorems.

In this chapter, such results are proven. These results are then used in the following chapter (boolean contractions).

1. Eigenelements of a Boolean Matrix

The set of the 2^n boolean vectors having n components is denoted by $\{0, 1\}^n$. A *boolean matrix* is furthermore understood to be an $n \times n$ matrix with elements from $\{0, 1\}$. Let B now be a boolean matrix.

Definitions. An element u (not equal to the zero vector) from $\{0, 1\}^n$ is called *a (boolean) eigenvector* of B if there exists $\lambda \in \{0, 1\}$ such that

$$B u = \lambda u \quad \text{(boolean operations).}$$

λ will be called *the (boolean) eigenvalue* associated with the eigenvector u.

Clearly, if a boolean matrix has an eigenvalue then it can only be 0 or 1.

Examples. (1) All diagonal boolean matrices D have the unit vectors

$$e_i = \begin{bmatrix} 0 \\ \vdots \\ 1 \\ \vdots \\ 0 \end{bmatrix} \leftarrow i \qquad (i=1,2,\ldots,n)$$

as eigenvectors, with the diagonal elements of D being the eigenvalues.

(2) The boolean matrix $\begin{bmatrix} 0 & 1 & 1 \\ 1 & 0 & 0 \\ 0 & 0 & 0 \end{bmatrix}$ has an eigenvalue 1 associated with the eigenvector $\begin{bmatrix} 1 \\ 1 \\ 0 \end{bmatrix}$, *but it does not have the eigenvalue* 0. (Be careful!)

Remark. If B has an eigenvalue λ with eigenvector u, then for all permutation matrices P, the boolean matrix $P^t BP$ still has the eigenvalue λ with eigenvector $v = P^t u$ such that

$$Bu = \lambda u \quad \text{from which} \quad P^t BPP^t u = \lambda P^t u.$$

Thus B and $P^t BP$ have the same eigenvalues ($B = P(P^t BP) P^t$).

In the following one will study and characterize the eigenvalues of an arbitrary boolean matrix.

Theorem 1. *A necessary and sufficient condition for a boolean matrix B to have a zero eigenvalue is that B has one or more zero columns. If B does have a zero eigenvalue then the associated eigenvectors are found by forming boolean sums of basis vectors corresponding to the zero columns of B.*

Indeed: If the i-the column of B is zero, then clearly $Be_i = 0 = 0.\ e_i$ which means that e_i is an eigenvector of B associated with the eigenvalue 0.

Conversely, it is clear that, if there is a non-zero vector u such that $Bu = 0$, then for all components u_i which are non-zero (that is, equal to 1) of u, the i-th column of B *must* be zero.

If I is a subset of $\{1, 2, \ldots, n\}$ such that the columns of B indexed by members of I are zero, then clearly $B \sum_{i \in I} e_i = \sum_{i \in I} Be_i = 0$ from which it follows that $u = \sum_{i \in I} e_i$ is an eigenvector of B associated with the eigenvalue 0.

Conversely if there is $u \neq 0$ such that $Bu = 0$, then u can be written $u = \sum_{i \in I} e_i$ where I is the set of indices of the non-zero components of u, and for these indices it follows that the corresponding columns of B are necessarily zero. □

We will now establish a *standard form* for an arbitrary (square) boolean matrix. This form (*which can be calculated in a finite number of operations*, as opposed to the Jordan normal form for an arbitrary real matrix) will give us complete information on the structure of the eigenelements.

Theorem 2. *Let B be a boolean matrix. Then either*

1) B has no zero rows or

2) there exists a permutation matrix P such that $P^t BP$ has the form

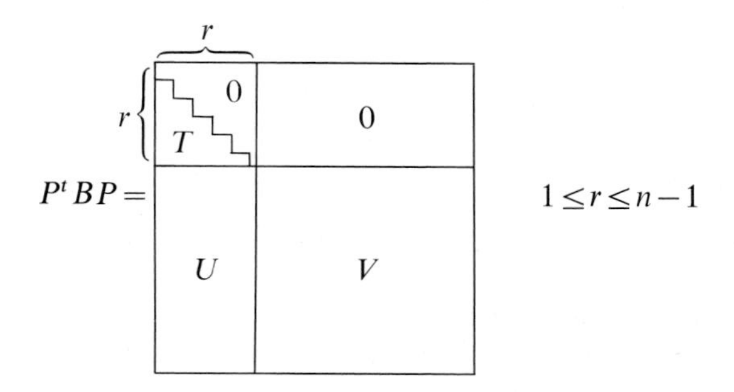

where T is strictly lower triangular and where V has no zero rows or

3) there exists a permutation matrix P such that $P^t BP = T$ and such that T is strictly lower triangular.

Definition. The matrix $P^t BP$ of point 2) is called *the standard form* of B.

This standard form clearly has the limiting cases of B itself (point 1) and T (point 3).

Proof of Theorem 2. First of all it may be that B has no zero rows and point 1 is satisfied.

If, however, B has one or more zero rows then one of these rows may be shifted into the first row by a certain permutation of rows. We also perform the same permutation on the columns (the row and the column permutations together result in an operation of the form $P_1^t BP_1$): obtaining

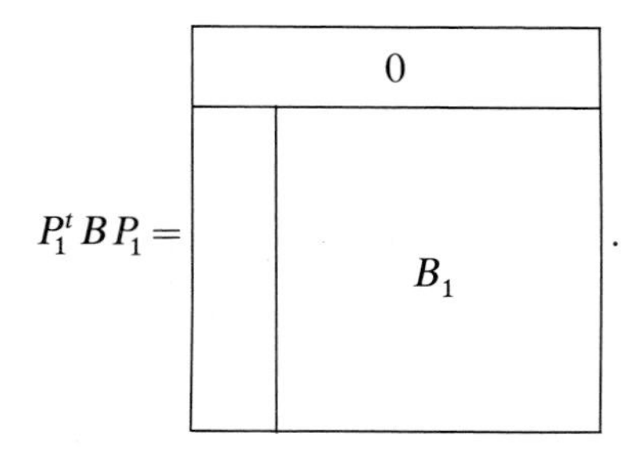

One may now proceed in the same manner with B_1 and so on. In the end the process may stop at step $r<n$. One therefore has $P_1, P_2, \ldots, P_r$ such that

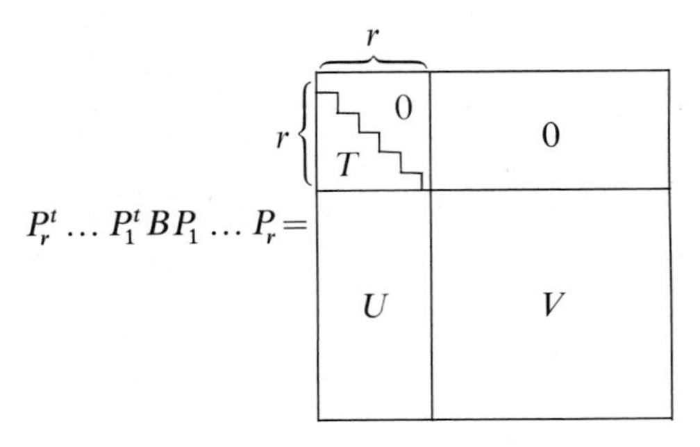

where T is strictly lower triangular and where V has no zero rows. Letting $P=P_1 P_2 \ldots P_r$ one obtains the result of point 2.

If the above procedure may be carried out until $r=n$ then one obtains point 3. The points 1) and 3) are clearly the limiting cases of point 2). □

Remark. The eigenvalues of B and $P^t BP$ are clearly the same (see p. 44).

Example.

$$B=\begin{bmatrix} 0 & 0 & 0 & 0 & 0 & 0 \\ 1 & 0 & 1 & 0 & 0 & 0 \\ 1 & 0 & 0 & 0 & 0 & 0 \\ 1 & 0 & 1 & 1 & 1 & 1 \\ 1 & 0 & 1 & 0 & 1 & 1 \\ 0 & 0 & 0 & 0 & 0 & 0 \end{bmatrix}.$$

Without doing the permutations one may

– cross out those rows of B that contain only zeros as well as the columns having the same indices

– iterate the procedure until it stops.

It goes as follows:

$$\begin{bmatrix} 0 & 0 & 0 & 0 & 0 & 0 \\ 1 & 0 & 1 & 0 & 0 & 0 \\ 1 & 0 & 0 & 0 & 0 & 0 \\ 1 & 0 & 1 & 1 & 1 & 1 \\ 1 & 0 & 1 & 0 & 1 & 1 \\ 0 & 0 & 0 & 0 & 0 & 0 \end{bmatrix} \begin{matrix} 1 \\ 4 \\ 3 \\ \\ \\ 2 \end{matrix}.$$

This process stops when $r=4$, and we have

$$P=P_1 P_2 P_3 P_4$$

$$=\underset{P_2}{\begin{bmatrix}1&0&0&0&0&0\\0&0&0&0&0&1\\0&0&1&0&0&0\\0&0&0&1&0&0\\0&0&0&0&1&0\\0&1&0&0&0&0\end{bmatrix}}\cdot\underset{P_4}{\begin{bmatrix}1&0&0&0&0&0\\0&1&0&0&0&0\\0&0&1&0&0&0\\0&0&0&0&0&1\\0&0&0&0&1&0\\0&0&0&1&0&0\end{bmatrix}}=\begin{bmatrix}1&0&0&0&0&0\\0&0&0&1&0&0\\0&0&1&0&0&0\\0&0&0&0&0&1\\0&0&0&0&1&0\\0&1&0&0&0&0\end{bmatrix}$$

since $P_1=P_3=I$ (unit matrix).

The standard form of B is then obtained as

$$P^t BP=\left[\begin{array}{cccc|c|c}0&0&0&0&0&0\\0&0&0&0&0&0\\1&0&0&0&0&0\\1&0&1&0&0&0\\\hline 1&1&1&0&1&0\\1&1&1&0&1&1\end{array}\right].$$

Theorem 3. *For a boolean matrix B to have the eigenvalue 1 it is necessary and sufficient that it contains a principal sub-matrix having no zero rows.*

To say that B contains a principal sub-matrix that has no zero rows is the same as to say that one has cases 1) or 2) of the preceding theorem. We now show that in cases 1) and 2) of that theorem, B has an eigenvalue 1, whereas in case 3) it does not.

– In the case 1) the vector e whose components are all equal to 1 clearly satisfies $Be=e$. 1 is therefore an eigenvalue of B.

– In the case 2) one has

$$\left[\begin{array}{c|c}T & 0\\\hline U & V\end{array}\right]\begin{bmatrix}0\\\hline 1\\1\\1\\\vdots\\1\end{bmatrix}=\begin{bmatrix}0\\\hline 1\\1\\1\\\vdots\\1\end{bmatrix}$$

where V has no zero rows. Then 1 is an eigenvalue of $P^t BP$ and hence of B.

– In the case 3), let u be such that $Tu=u$ where T is strictly triangular. One then obtains

$$u_1=0 \quad \text{then} \quad u_2=0 \quad \text{then} \quad \dots \; u_n=0.$$

However, $u=0$ may not be an eigenvector. □

Remark. It is possible to characterize (in a somewhat technical manner) the eigenvectors associated with the eigenvalue 1.

Theorem 4. *All boolean matrices have an eigenvalue.*

This result is not a priori obvious. It is, however, an elementary consequence of Theorems 1, 2 and 3.

Let us therefore consider cases 1) and 2) of Theorem 2. Then, according to Theorem 3, B has an eigenvalue 1.

In the case 3) B does not have the eigenvalue 1 (all principal submatrices of B have at least one zero row), but B then has at least one zero column which means that it has the eigenvalue 0 according to Theorem 1. □

Examples.

$\begin{bmatrix} 1 & 1 \\ 1 & 1 \end{bmatrix}$ has eigenvalue 1, *but not an eigenvalue* 0.

$\begin{bmatrix} 1 & 0 \\ 1 & 1 \end{bmatrix}$ the same

$\begin{bmatrix} 0 & 0 \\ 1 & 1 \end{bmatrix}$ *the same*

$\begin{bmatrix} 0 & 0 \\ 0 & 1 \end{bmatrix}$ has eigenvalues 1 and 0

$\begin{bmatrix} 0 & 0 \\ 1 & 0 \end{bmatrix}$ has eigenvalue 0, but not the eigenvalue 1.

Definition. $\rho(B)$ will denote "the largest boolean eigenvalue" of B ($\rho(B)=0$ or 1) and it will be called the *(boolean) spectral radius* of B.

Theorem 3 above gives therefore a necessary and sufficient condition for $\rho(B)=1$. Moreover we have

Theorem 5. *Let B be an $n\times n$ boolean matrix. In order that $\rho(B)=0$, it is necessary and sufficient that one of the following (equivalent) conditions are satisfied:*

a) there exists a permutation matrix P such that $P^t BP$ is strictly lower triangular.

b) There exists an integer $p \leq n$ *such that*

$$B^p = 0 \qquad (\textit{boolean power})$$

[*which, of course, also implies* $B^n = 0$].

Point a) is obvious given the previous theorem. It is case 3) of Theorem 2.

We now show point b):

Clearly if $\rho(B) = 0$ then there exists a permutation matrix P such that $P^t BP$ is *strictly* lower triangular. This means there exists an integer $p \leq n$ such that

$$(P^t BP)^p = 0 \qquad \text{(boolean power)}$$

and therefore

$$P^t B^p P = 0$$

and finally

$$B^p = 0.$$

Conversely, if $\rho(B) \neq 0$ then B has an eigenvalue 1. That is there exists $u \neq 0$ such that $Bu = u$, which implies that for all integer p, $B^p u = u$. Hence B^p can not be the zero matrix. □

Remarks. (1) The result given in Theorem 5 above is formally the same as the result for matrices with positive or zero real elements, when the usual spectral radius is considered (see [44]).

(2) It is clear that in cases 2) and 3) of Theorem 2, the matrix B is *reducible**.

If a boolean matrix is irreducible, then it satisfies case 1), it has in particular $\rho(B) = 1$ and the vector e having all components equal to 1 is an eigenvector of B associated with the eigenvalue 1.

Conversely, a matrix may satisfy case 1) and be reducible

$$\begin{bmatrix} 1 & 0 \\ 1 & 1 \end{bmatrix}$$

and also have the eigenvalue 0:

$$\begin{bmatrix} 1 & 0 \\ 1 & 0 \end{bmatrix}.$$

(3) Finally, in order that a boolean matrix will have the spectral radius equal to 1 it is necessary and sufficient that points a) and b) of the

---

* That is to say that there exists a permutation matrix P such that $P^t BP$ is of the following form: $\begin{array}{|c|c|} \hline & 0 \\ \hline & \\ \hline \end{array}$ (diagonal blocks are square).

preceding theorem are not satisfied, that is B contains a principal submatrix that has no zero row. These are the cases 1) or 2) of Theorem 2.

Theorem 6. *Let B be an $n \times n$ boolean matrix.*

In order that B may have an eigenvalue 1 but not an eigenvalue 0, it is necessary and sufficient that it has no zero columns (which therefore implies that it has an eigenvalue 1).

The preceding Theorem 5 gives conditions for B to have an eigenvalue 0 but not 1 (necessary and sufficient conditions for $\rho(B)=0$).

From Theorem 1 we know that B having no zero columns is equivalent to the fact that B has no eigenvalue 0. Since B necessarily has an eigenvalue (Theorem 4) it follows that the last statement (B has no eigenvalue 0) is equivalent to B having the eigenvalue 1, but not the eigenvalue 0. This proves the first part of the theorem. □

Theorem 7. *Let B be an $n \times n$ boolean matrix and B^p the p-th boolean power ($p \geq 1$ an arbitrary integer).*

Then $\rho(B^p)=\rho(B)$.

If $\rho(B)=0$ then $B^n=0$ (Theorem 5) from which $B^{np}=0$ and then $\rho(B^p)=0$.

If $\rho(B)=1$ then there exists a non-zero vector u such that $Bu=u$ from which $B^p u=u$ for all p and $\rho(B^p)=1$. □

Remarks. (1) The above results are stated in an analogous manner for real matrices. In the boolean context, however, the proofs are elementary.

(2) *Be careful, however, in the case of blocks.* If

$$B=\left[\begin{array}{c|c} C & 0 \\ \hline E & D \end{array}\right] \quad \text{(square blocks on diagonal)}$$

then the eigenvalues of D are eigenvalues of B, *but the eigenvalues of C are not necessarily eigenvalues of B.*

Example.

$$B=\left[\begin{array}{cc|c} 0 & 0 & 0 \\ 1 & 0 & 0 \\ \hline 1 & 1 & 1 \end{array}\right]$$

B has only the eigenvalue 1 (no zero columns) which is also the eigenvalue of D, but C has a zero eigenvalue which is not an eigenvalue for B.

2. The Boolean Perron-Frobenius Theorem

We will now transpose the classical theorem of Perron-Frobenius [24], [25], [38], [44] into a boolean context. In the boolean context the proofs are much simpler and certain points of the theorem are trivial. These points will still be elaborated, however, in order to keep the analogy with the real version.

Two order relations are used, both denoted by $\leq$.

The inequality $\leq$ between two boolean vectors x and y is the "componentwise" inequality obtained from the order relation $0\leq 0\leq 1\leq 1$ in $\{0,1\}$.

The inequality $A\leq B$ between two boolean matrices is similarly an "elementwise" inequality

$$a_{ij}\leq b_{ij} \qquad (i,j=1,2,\ldots,n).$$

With these notations, here is the theorem:

The Boolean Perron-Frobenius Theorem. *Let B be an $n\times n$ boolean matrix. Then the following is true:*

1) Its spectral radius $\rho(B)$ is an eigenvalue, with an associated eigenvector $u\geq 0$.

2) If $B\leq C$ then $\rho(B)\leq\rho(C)$.

3) For all non-empty subsets M of $\{1,2,\ldots,n\}$ one has

$$\operatorname*{Max}_{M}\operatorname*{Min}_{i\in M}\sum_{j\in M} b_{ij}=\rho(B)\leq \operatorname*{Max}_{i=1,2,\ldots,n}\sum_{j=1}^{n} b_{ij} \qquad (\textit{boolean } \textstyle\sum).$$

In these relations the inequality $\leq$ is in general not an equality.

4) If B is irreducible, then $\rho(B)=1$, and there is (at least) a corresponding eigenvector e (with all components equal to 1) for which the equality

$$\rho(B)=\operatorname*{Max}_{i=1,2,\ldots,n}\sum_{j=1}^{n} b_{ij}=1$$

is valid.

5) For all $\lambda\in\{0,1\}$ such that $Be\leq\lambda e$ one has $\rho(B)\leq\lambda$.

6) Let $u\neq 0$. Then for all λ such that $\lambda u\leq Bu$ one has $\lambda\leq\rho(B)$.

Proof of Point 1). This point is trivial from the preceding discussions. It is only included because of the analogy with the classical theorem.

Proof of Point 2). Let C be a boolean matrix satisfying $B\leq C$. If $\rho(B)=1$ then B has a principal submatrix having no zero row and hence C as well which implies $\rho(C)=1$ (Theorem 3). If $\rho(B)=0$ then there is nothing to prove.

Proof of Point 3).

a) $\operatorname*{Max}_{i=1,2,\ldots,n} \sum_{j=1}^{n} b_{ij}$ (boolean $\sum$) is only zero if $B=0$, in which case $\rho(B)=0$. This proves the inequality $\rho(B) \leq \operatorname*{Max}_{i=1,2,\ldots,n} \sum_{j=1}^{n} b_{ij}$. The example $B = \begin{bmatrix} 0 & 0 \\ 1 & 0 \end{bmatrix}$ where $B \neq 0$, but $\rho(B)=0$ shows that the inequality is not in general an equality.

b) Let M be a non-empty subset of $\{1, 2, \ldots, n\}$. Then suppose

$$\operatorname*{Min}_{i \in M} \sum_{j \in M} b_{ij} = 1$$

which is the same as to say that the principal submatrix defined by the rows and columns of B indexed by M has no zero row.

If

$$\operatorname*{Max}_{M} \operatorname*{Min}_{i \in M} \sum_{j \in M} b_{ij} = 1,$$

then this means that B has a principal submatrix without a zero row. From Theorem 3 it follows that 1 is an eigenvalue of B and $\rho(B)=1$.

If, however,

$$\operatorname*{Max}_{M} \operatorname*{Min}_{i \in M} \sum_{j \in M} b_{ij} = 0$$

then this means that each principal submatrix of B has at least one zero row. According to Theorem 3 it follows that 1 is therefore not an eigenvalue which means that (Theorem 4) 0 is the only eigenvalue, from which $\rho(B)=0$.

Remark. For the matrix $B = \begin{bmatrix} 0 & 0 \\ 0 & 1 \end{bmatrix}$ one has $\rho(B)=1$ and

$$\operatorname*{Min}_{M} \operatorname*{Max}_{i \in M} \sum_{j \in M} b_{ij} = 0.$$

Proof of Points 4) and 5). Completely trivial after the preceding discussions, but mentioned because of the analogy with the standard version of the Perron-Frobenius theorem.

Proof of Point 6). Trivial for $\lambda=0$. For $\lambda=1$ let $u \neq 0$ be such that $Bu \geq u$. Then after a permutation of rows and columns one arrives at the following situation:

$$\begin{bmatrix} B_1 & B_2 \\ B_3 & B_4 \end{bmatrix} \begin{bmatrix} 0 \\ 1 \\ \vdots \\ 1 \end{bmatrix} = \begin{bmatrix} B_2 \\ B_4 \end{bmatrix} \begin{bmatrix} 1 \\ \vdots \\ 1 \end{bmatrix} \geq \begin{bmatrix} 0 \\ 1 \\ \vdots \\ 1 \end{bmatrix}$$

This means that $B_4 \begin{bmatrix} 1 \\ \vdots \\ 1 \end{bmatrix} \geq \begin{bmatrix} 1 \\ \vdots \\ 1 \end{bmatrix}$, that is to say $B_4 \begin{bmatrix} 1 \\ \vdots \\ 1 \end{bmatrix} = \begin{bmatrix} 1 \\ \vdots \\ 1 \end{bmatrix}$ and B_4 (a principal submatrix of B) does not have a zero row. Finally (Theorem 3) $\rho(B) = 1$. □

We now turn to:

3. The Boolean Stein-Rosenberg Theorems

The "truncated" Stein-Rosenberg Theorem. *Let L and U be $n \times n$ boolean matrices. If one defines*

$$S_r = [I + L + \ldots + L^{r-1}]\, U + L^r \qquad (r = 1, 2, \ldots)$$

then all the matrices S_r $(r \geq 1)$ have the same boolean spectral radius.

Indeed, the matrices S_r may be defined by the recurrence relation $(S_0 = I)$

$$S_{r+1} = L S_r + U \qquad (r = 0, 1, 2, \ldots).$$

a) Let us first assume that there exists a $u \neq 0$ such that $S_1 u = u$ $(\rho(S_1) = 1)$. Then $S_r u = u$ since if we assume $S_{r-1} u = u$ we get

$$S_r u = L S_{r-1} u + U u = L u + U u = S_1 u = u \qquad (r = 2, 3, \ldots)$$

from which $\rho(S_r) = 1$ for all $r \geq 1$.

b) Let us then assume for the contrary that $\rho(S_1) = 0$. Then (Theorem 5) there exists a permutation matrix P such that $\bar{S}_1 = P^t S_1 P$ is strictly lower triangular.

Now we have $S_1 = L + U$ from which

$$\bar{S}_1 = P^t L P + P^t U P$$

and $\bar{L} = P^t L P \leq \bar{S}_1$ which implies that $\bar{L}$ is strictly lower triangular, and $\bar{U} = P^t U P \leq \bar{S}_1$ which implies that $\bar{U}$ is strictly lower triangular.

Now, since $S_r = [I + L + \ldots + L^{r-1}]\, U + L^r$ $(r \geq 1)$ it is easy to see that the matrix $\bar{S}_r = P^t S_r P$ can be written as

$$\bar{S}_r = [I + \bar{L} + \ldots + \bar{L}^{r-1}]\, \bar{U} + \bar{L}^r$$

from which it follows that $\bar{S}_r$ is *strictly lower triangular.**

* $[I + \bar{L} + \ldots + \bar{L}^{r-1}]$ is lower triangular with unit diagonal and its product with $\bar{U}$ (strictly lower triangular) results in a strictly lower triangular matrix. This matrix added to a strictly lower triangular $\bar{L}^r$ results in $\bar{S}_r$ having the required form.

Therefore $\rho(\bar{S}_r)=0$. Now $S_r = P\bar{S}_r P^t$ from which (see Remark, p. 44) $\rho(S_r) = \rho(\bar{S}_r)=0$. □

We will now, moreover, give another proof of point b), which completes the preceding theorem in a useful manner.

Corollary. *If $\rho(S_1)=0$, then there exists (see Theorem 5) an integer $p \leq n$ such that*

$$S_1^p = 0 \qquad (\textit{boolean power}).$$

Then for all $r \geq 1$ one again has

$$S_r^p = 0.$$

We have $S_1 = L+U$. Developing now $(L+U)^p$ one obtains a sum of terms of the form $A_1 A_2 \dots A_p$ where A_i is either L or U $(i=1,2,\dots,p)$.

This (boolean) sum of matrices is zero if and only if each of the terms is zero. Then

$$A_1 \dots A_p = 0 \qquad \text{where } A_i \text{ may be either } L \text{ or } U \ (i=1,2,\dots,p).$$

We therefore expand $(S_r)^p = [(I+L+\dots+L^{r-1})\,U + L^r]^p$. This contains nothing but terms of the preceding type which implies $(S_r)^p = 0$ and therefore $\rho(S_r)=0$. □

This corollary is very useful in order to prove (the following chapter) that if an operator F is contracting, *then the associated Gauss-Seidel operator is more contracting.*

Remarks. (1) In the preceding theorem one evidently has $L \leq L+U=S$ from which (Perron-Frobenius) $\rho(L) \leq \rho(S_1)$. Whenever $\rho(S_1)=0$ one therefore has $\rho(L)=0$ (moreover, $\bar{L}=P^t LP$ is strictly lower triangular). Then $L^n=0$ which means that, starting from $r=n$, S_r stabilizes at the matrix

$$[I+L+\dots+L^{n-1}]\,U$$

which then in particular has the spectral radius zero.

(2) Whenever $\rho(S_1)=1$ with $\rho(L)=0$, one still has $L^n=0$ so that, starting from $r=n$, S_r is stabilized at the matrix $[I+L+\dots+L^{n-1}]\,U$ which then has the spectral radius 1.

With the above results one has the

Boolean Stein-Rosenberg Theorem. *Let L and U be two $n \times n$ boolean matrices.*

a) If $\rho(L+U)=0$ then $\rho((I+L+\dots+L^{n-1})\,U)=0$. More precisely: there exists an integer $p \leq n$ such that

$$(L+U)^p = 0$$

and for the same p

$$[(I+L+\ldots+L^{n-1})\,U]^p=0.$$

$$b)\ \textit{If}\ \begin{cases}\rho(L+U)=1\\ \rho(L)=0\end{cases}\ \textit{then}\ \rho((I+L+\ldots+L^{n-1})\,U)=1.$$

Remarks. (1) The condition $\rho(L)=0$ may not be relaxed in case b), because with L being such that $\rho(L)=1$ and with U being the zero matrix it is clear that

$$\rho((I+L+\ldots+L^{n-1})\,U)=\rho(0)=0$$

and still $\rho(L+U)=\rho(L)=1$.

(2) One notes the clear analogy of all the results here with the classical Stein-Rosenberg theorem [25], [38], [44] and with the so-called truncated Stein-Rosenberg theorem [36], [38] valid for square matrices with *real* elements ≥ 0.

4. Conclusion

The theorems of Perron-Frobenius and Stein-Rosenberg "fit" perfectly into the context of boolean matrices with the differences that the proofs are much simpler and the arguments used in the proofs are *finite* arguments (the latter is the main difference between this case and the case with matrices with real elements).

In the next chapter we make a rather strong assumption (carried by the incidence matrix $B(F)$) regarding the *contraction* of an operator F relative to the discrete metric tool d. With this assumption we use the results of the present chapter to study and compare corresponding serial and parallel iterations for F on X.

4. Boolean Contraction and Applications

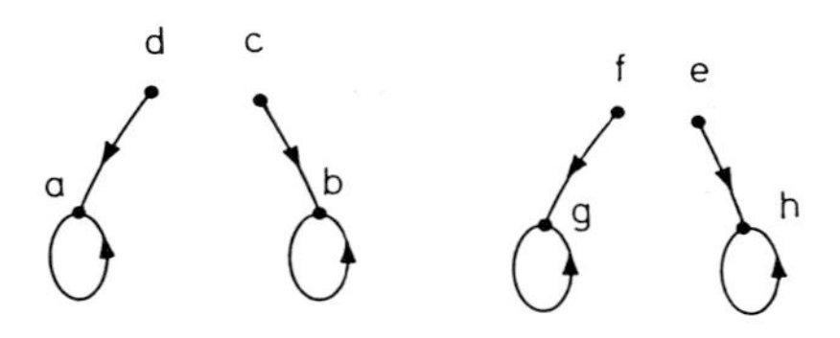

Again, let $X=\prod_{i=1}^{n} X_i$ denote the product of n finite sets X_i and let F be a map of X into itself. We are interested in the behaviour of the associated finite automata network (see Chap. 1) when it operates in a parallel mode (successive approximations to F), in a serial mode (Gauss-Seidel to F) and in a serial-parallel mode.

A simple condition will be established in this chapter which will assure *the existence and the unicity of a fixed point of* F (stable configuration). Under this condition, the fixed point of F *is reached in no more than n steps independent of both the operating mode and the starting configuration.* This sufficient condition *(contraction of F relative to d)* will then both assure that the iteration graph for F is *simple* and therefore the simplicity of the graph of G, as well as the simplicity of the graphs of all the block Gauss-Seidel operators (associated with given serial-parallel modes). *Furthermore, when one passes from a given operating mode to a more sequential operating mode* (for example from F to G), *the contraction is strengthened, that is to say, fewer steps are required for convergence.*

Let us recall the basic inequality (Theorem 1, Chap. 2)

$$\forall x, y \in X \qquad d(F(x), F(y)) \leq B(F)\, d(x, y)$$

where $B(F)$ is the incidence matrix of F.

The contraction of F relative to the discrete metric δ on X is (see introduction to Chap. 2) a sufficient condition for the iteration graph for F to have only one basin with one fixed point (that is to say, the condition for simplicity). In effect, this imposes the condition that F is constant.

The contraction of F relative to the vector distance d is much less restrictive (but still strong enough). It will still, in particular, guarantee the simplicity of the iteration graph for F.

1. Boolean Contraction

Definition. F is said to be *contracting* relative to the vector distance d if there exists an $n \times n$ boolean matrix M having spectral radius zero, such that

$$\forall x, y \in X \qquad d(F(x), F(y)) \leq M\, d(x, y).$$

The following characterizations are now valid:

Theorem 1. *F is contracting relative to the vector distance d if and only if one of the four equivalent conditions are valid:*

1) $\rho(B(F))=0$ (boolean spectral radius).

2) There exists an integer $p \leq n$ such that $[B(F)]^p=0$ (boolean power).*

3) There exists an $n \times n$ permutation matrix P such that

$$P^t B(F) P$$

is strictly lower triangular.

4) The connectivity graph for F (the graph of the associated automata network) has no circuits (a sequence of oriented arcs that return to their vertex of departure).

First of all, the results of the preceding chapter, Theorem 5, show that points 1), 2) and 3) are equivalent. Using 3) it is furthermore easy to see that they are also equivalent to 4). We now show that points 1), 2) and 3) are moreover equivalent to the contraction of F relative to d.

If $\rho(B(F))=0$ then F is clearly contracting (take $M=B(F)$).

Conversely, if there exists a boolean matrix M guaranteeing the inequality for the contraction, then according to Theorem 2 of Chap. 2 we have

$$B(F) \leq M$$

and using the boolean Perron-Frobenius Theorem we get

$$0 \leq \rho(B(F)) \leq \rho(M) = 0$$

which implies the result. □

2. A Fixed Point Theorem

We will now establish a fixed point theorem.

Theorem 2. *If F is contracting relative to d then there exists an integer $p \leq n$ such that F^p is constant. Stated differently, there exists a $\xi \in X$ such that, $\forall x \in X$, $F^p(x)=\xi$.*

---

* From which, clearly $[B(F)]^n=0$ (boolean power).

This ξ is the unique fixed point (stable configuration) of F in X, and for all x^0 in X, the iteration $x^{r+1}=F(x^r)$ is stationary at ξ for $r=p, p+1, \ldots$.

Since $\rho(B(F))=0$ there exists an integer $p \leq n$ such that

$$[B(F)]^p=0 \qquad \text{(boolean power)}$$

from which $B(F^p) \leq [B(F)]^p=0$.

This means that $B(F^p)=0$, which implies that $F^p(x)$ does not depend on x. Thus F^p is constant and there therefore exists ξ in X such that

$$\forall x \in X \qquad F^p(x)=\xi.$$

This proves the first point.

Furthermore, we then have

$$F^{p+1}(\xi)=F^p(F(\xi))=\xi=F(F^p(\xi))=F(\xi)$$

from which $\xi=F(\xi)$, that is, ξ is a fixed point for F. ξ is clearly unique since if η is another fixed point of F, then

$$\xi=F^p(\xi)=F^p(\eta)=\eta$$

which completes the proof of the theorem. □

Remarks. (1) From Theorem 1, point 3) it follows that if F is contracting relative to d then using a convenient numbering of the components f_i and the same numbering for the variables x_j it is possible to write F in the following form (example for $n=3$):

$$\begin{aligned} &f_1(\sqcup, \sqcup, \sqcup) \qquad (\forall x \ f_1(x)=\text{constant}) \\ &f_2(x_1, \sqcup, \sqcup) \\ &f_3(x_1, x_2, \sqcup). \end{aligned}$$

(This notation means that f_1 does not depend on any component of x, f_2 only on x_1 and f_3 on x_1 and x_2, but not on x_3.)

The evolution of the *parallel* iteration for F is not affected by the change in numbering. Therefore, the results of Theorem 2 appear clearly in this form. Indeed, the components of the fixed point ξ are

$$\begin{aligned} &\xi_1=f_1(\sqcup, \sqcup, \sqcup) \\ &\xi_2=f_2(\xi_1, \sqcup, \sqcup) \\ &\xi_3=f_3(\xi_1, \xi_2, \sqcup). \end{aligned}$$

Moreover, the iteration on F, starting at x^0, evolves in the following manner:

$$x^1\begin{cases} x_1^1 = f_1(\sqcup, \sqcup, \sqcup) = \xi_1 \\ x_2^1 = f_2(x_1^0, \sqcup, \sqcup) \\ x_3^1 = f_3(x_1^0, x_2^0, \sqcup) \end{cases}$$

$$x^2\begin{cases} x_1^2 = f_1(\sqcup, \sqcup, \sqcup) = \xi_1 \\ x_2^2 = f_2(\xi_1, \sqcup, \sqcup) = \xi_2 \\ x_3^2 = f_3(\xi_1, x_2^1, \sqcup) \end{cases}$$

$$x^3\begin{cases} x_1^3 = f_1(\sqcup, \sqcup, \sqcup) = \xi_1 \\ x_2^3 = f_2(\xi_1, \sqcup, \sqcup) = \xi_2 \\ x_3^3 = f_3(\xi_1, \xi_2, \sqcup) = \xi_3. \end{cases} \quad \text{from which } x^3 = \xi.$$

(2) The contraction of F relative to d guarantees that the iteration graph for F contains only one basin which contains only one fixed point. This is the same as to say that the iteration graph for F is simple. Furthermore, *at most n steps* are required to "descend" from any point x^0 in X to the fixed point of F.

(3) If F is contracting relative to d then there exists $p \leq n$ such that F^p is constant (and therefore contracting relative to the discrete metric δ).

Thus the contraction of F relative to d implies that F is the p-th power-contractant relative to the discrete metric δ on X. One may in this manner also verify the result obtained above.

(4) The result established above may be interpreted in the following manner:

If the connectivity graph for F (that is, the graph of the associated automata network) *does not have a circuit then the iteration graph for F is simple.*

The following interesting question is posed: *What properties of the iteration graph for F* (which contains 2^n points in the simple (current) case where $X = \{0, 1\}$) *can be given only based on the information provided by the connectivity graph for F* (which has n points)? It turns out that we just gave a partial answer to this question in the previous discussion.

(5) Clearly, to a given connectivity graph, there corresponds many F (one has already stated that the information given by $B(F)$ is relatively rudimentary).

If the spectral radius of $B(F)$ is zero (the connectivity graph has no circuits) then *all* the functions F having this $B(F)$ as their incidence matrix have a simple iteration graph (see Example 1 below).

(6) Evidently, the contraction of F is a *sufficient*, but not *necessary* condition for the iteration graph of F to be simple (see Example 3 below).

(7) *Condensation.* It is clearly possible to condense d into a "more compact" vector distance γ (see Chap. 2, p. 40) and to define the contraction of F relative to γ. One then obtains a *more restrictive condition* for the contraction compared to the condition relative to d. In order to give a

feeling for this we turn to the following example where we use the distance γ already defined for $X=\{0,1\}^6$ on p. 40 by

$$\gamma(x,y)=\begin{bmatrix}\operatorname{Max}(\delta_1(x_1,y_1),\delta_2(x_2,y_2)\,\delta_3(x_3,y_3))\\ \operatorname{Max}(\delta_4(x_4,y_4),\delta_5(x_5,y_5))\\ \delta_6(x_6,y_6)\end{bmatrix}.$$

Let F: $\{0,1\}^6\to\{0,1\}^6$ such that

$$\gamma(F(x),F(y))\le M\,\gamma(x,y)$$

with

$$M=\begin{bmatrix}0&1&1\\0&0&0\\0&1&0\end{bmatrix}.$$

The boolean spectral radius of M is zero, which means that F is contracting relative to γ. From this, one has a result analogous to that established in Theorem 2.

However, the form of M implies that $B(F)$ has at least the following zeros:

$$B(F)=\left[\begin{array}{ccc|cc|c}0&0&0&x&x&x\\0&0&0&x&x&x\\0&0&0&x&x&x\\\hline 0&0&0&0&0&0\\0&0&0&0&0&0\\\hline 0&0&0&x&x&0\end{array}\right]\quad \text{the } x\text{'s may be either 0 or 1.}$$

This certainly implies that B(F) has the spectral radius zero, which in turn implies that F is contracting relative to d.

Conversely, however, F may be contracting relative to d, but not relative to γ. It is sufficient to consider the matrix

$$B(F)=\left[\begin{array}{ccc|cc|c}0&0&0&x&x&x\\1&0&0&x&x&x\\1&1&0&x&x&x\\\hline 0&0&0&0&0&0\\0&0&0&1&0&0\\\hline 0&0&0&x&x&0\end{array}\right].$$

In this example, $\rho(B(F))=0$ and therefore all F that have $B(F)$ as incidence matrix are contracting relative to d, but not relative to γ since

$$M \geq \begin{bmatrix} 1 & 0 & 0 \\ 0 & 1 & 0 \\ 0 & 0 & 0 \end{bmatrix} = N.$$

Therefore $\rho(M) \geq \rho(N) = 1$.

Therefore the contraction relative to d is the least restrictive of all the possible conditions for contraction that one may define on the basis of vector distances condensed from d. (This is true since any strictly *block* triangular matrix (square blocks on the diagonal) is strictly triangular. The converse is, however, false.)

3. Examples

1) $X = \{0, 1\}^3$ F is given by the table

	x			$F(x)$			
a	0	0	0	0	1	0	c
b	0	0	1	1	1	0	g
c	0	1	0	1	1	1	h
d	0	1	1	0	1	1	d
e	1	0	0	0	1	0	c
f	1	0	1	1	1	0	g
g	1	1	0	1	1	1	h
h	1	1	1	0	1	1	d

F may be represented by an automata network having 3 cells, each cell may take on two states (0 or 1). The transition function is the following:

$$f_1(x_1, x_2, x_3) = \bar{x}_2 x_3 + x_2 \bar{x}_3$$
$$f_2(x_1, x_2, x_3) = 1$$
$$f_3(x_1, x_2, x_3) = x_2.$$

From this we get the incidence matrix of F and its connectivity graph

$$B(F) = \begin{bmatrix} 0 & 1 & 1 \\ 0 & 0 & 0 \\ 0 & 1 & 0 \end{bmatrix}.$$

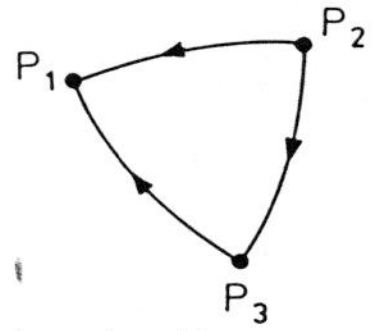

P_1 P_2 P_3

(graph without circuit)

One verifies that $\rho(B(F)) = 0$ from which it follows that F is contracting.

The iteration graph is the following:

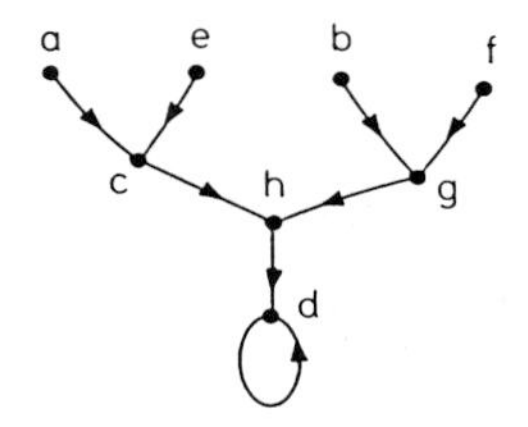

Indeed, this graph is simple (and d is the unique fixed point).

With the same connectivity graph, that is to say, the same incidence matrix, the function H given by

$$h_1(x_1, x_2, x_3) = x_2 + \bar{x}_3$$
$$h_2(x_1, x_2, x_3) = 0$$
$$h_3(x_1, x_2, x_3) = \bar{x}_2$$

has the following (simple) iteration graph:

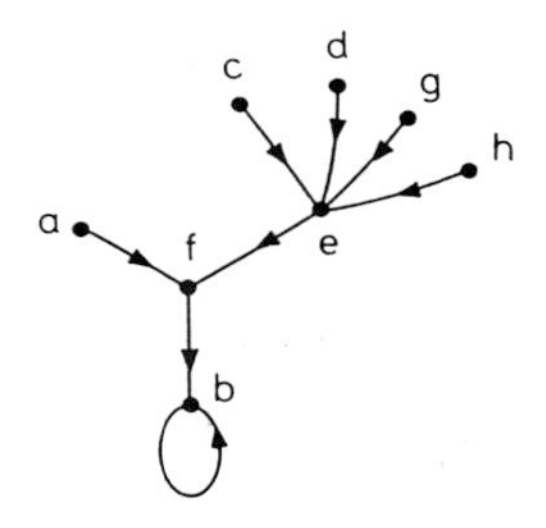

In these two graphs, the unique fixed point is reached in no more than three steps.

2) $X = \{0, 1\}^4 = \{a, b, c, \ldots, p\}$.

Let us take the incidence matrix $B(F)$ as

$$B(F) = \begin{bmatrix} 0 & 0 & 1 & 0 \\ 1 & 0 & 1 & 0 \\ 0 & 0 & 0 & 0 \\ 1 & 1 & 1 & 0 \end{bmatrix} \qquad \rho(B(F)) = 0$$

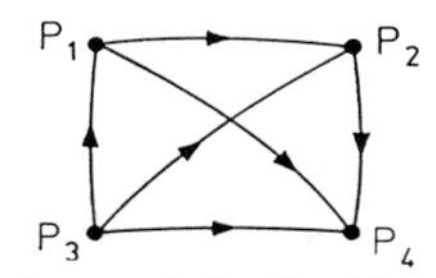

Connectivity graph for F, without circuit

and for example

$$f_1(x)=\bar{x}_3$$
$$f_2(x)=x_1 x_3$$
$$f_3(x)=1$$
$$f_4(x)=x_1+x_2 x_3.$$

The iteration graph for F is now obtained after some calculations as

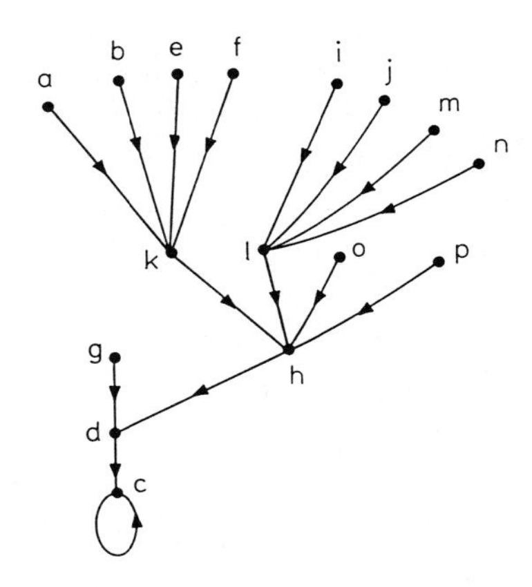

Clearly the graph is simple and furthermore the fixed point c is always reached in at most four steps.

3) $X=\{0, 1\}^3$ and F is defined by

$$f_1(x)=\bar{x}_1 x_2+x_1 x_2 \bar{x}_3$$
$$f_2(x)=0$$
$$f_3(x)=x_1+x_2$$

$$B(F)=\begin{bmatrix}1 & 1 & 1\\ 0 & 0 & 0\\ 1 & 1 & 0\end{bmatrix}$$

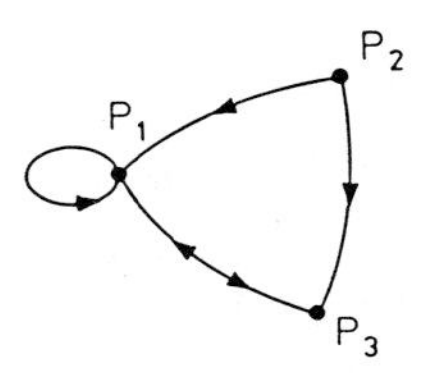

$\rho(B(F))=1$ and the connectivity graph contains one circuit (for example 1 3 1). F is therefore *not* contracting relative to d.

Even so, when all the calculations are done, the iteration graph for F is simple

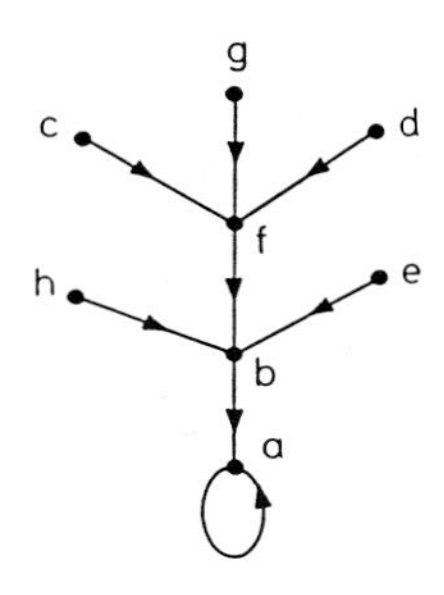

The contraction of F is therefore only a *sufficient* condition for the iteration graph of F to be simple.

4. Serial Mode: Gauss-Seidel Iteration for a Contracting Operator

We will now discuss a basic result. *If F is contracting then the associated Gauss-Seidel operator G is at least as contracting as F.* This means that *for a contracting operator, the serial iteration procedure proceeds at least as fast* (and in practice, much faster) *towards the unique fixed point as the parallel iteration procedure.*

Let us recall how one decomposed the incidence matrix $B(F)$ into L and U (L the strictly lower triangular, and U the upper triangular part of $B(F)$)

$$B(F)=\begin{bmatrix} & U \\ L & \end{bmatrix}.$$

Furthermore, one had the following result (Chap. 2, Theorem 6):

$$B(G)\leq(I+L+\ldots+L^{n-1})\,U.$$

Theorem 3. *If F is contracting relative to d, then there exists (see Theorem 2) an integer p such that*

$$[B(F)]^p=0 \qquad (p\leq n).$$

Therefore G is also contracting, and at least as much as F because

$$[B(G)]^p=0.$$

Then, in the same manner as for F, the iteration graph for G is simple. If the iteration is started from any x^0 in X, then the Gauss-Seidel method (serial process) becomes stationary at the unique fixed point of F (and G) in no more than p steps.

This result is based on the boolean Stein-Rosenberg theorem (Chap. 3, p. 54).

The spectral radius of $B(F)=L+U$ is zero. Therefore, let p be the smallest integer such that

$$[B(F)]^p=0 \qquad \text{(boolean power).}$$

Then, for that p we have

$$0 \leq [B(G)]^p \leq [(I+L+\ldots+L^{n-1})\,U]^p=0$$

from which

$$[B(G)]^p=0.$$

This furthermore means that $\rho(B(G))=0$, which implies that G is contracting relative to d. G then has a unique fixed point which is evidently that of F and the remainder of the theorem follows. □

Remark. The above theorem corresponds, in the context of discrete iterations developed here, to a similar result concerning the Gauss-Seidel method for non-linear systems with fixed points in $\mathbb{R}^n$ [32], [39]. The proofs differ, although they correspond in outline. Here the boolean Stein-Rosenberg theorem is used, whereas the standard Stein-Rosenberg theorem is used in [32], [39].

5. Examples

1) Let us reconsider the example 1) studied above. We have

$$F(x)=\begin{bmatrix} f_1(x)=\bar{x}_2 x_3+x_2\bar{x}_3 \\ f_2(x)=1 \\ f_3(x)=x_2 \end{bmatrix}$$

from which G is obtained as

$$G(x)=\begin{bmatrix} g_1(x)=\bar{x}_2 x_3+x_2\bar{x}_3 \\ g_2(x)=1 \\ g_3(x)=1. \end{bmatrix}$$

This results in

$$B(G)=\begin{bmatrix} 0 & 1 & 1 \\ 0 & 0 & 0 \\ 0 & 0 & 0 \end{bmatrix}.$$

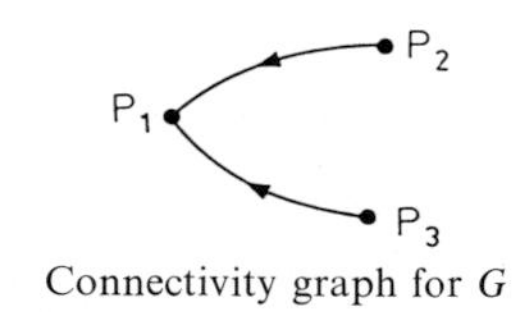

Connectivity graph for G

This graph does not have a circuit and one verifies that G is contracting. The iteration graph for G is furthermore simple.

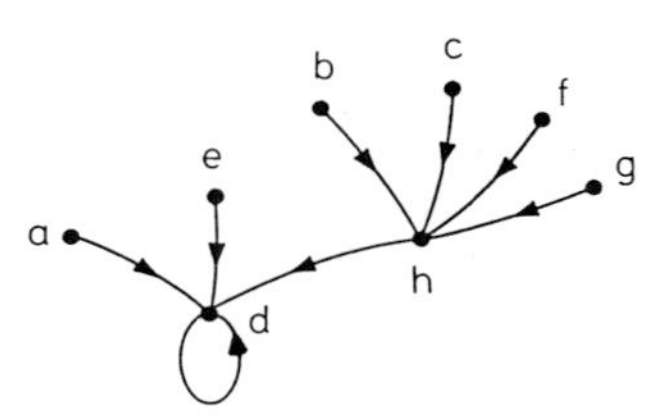

The convergence is attained in no more than *three* steps for F, whereas for G it is attained *in at most two steps*. G is therefore "more contracting" than F.

Remark. In this example, one has

$$B(F)=\begin{bmatrix}0 & 1 & 1\\ 0 & 0 & 0\\ 0 & 1 & 0\end{bmatrix},$$

from which

$$B(F_1)=\begin{bmatrix}0 & 1 & 1\\ 0 & 1 & 0\\ 0 & 0 & 1\end{bmatrix},\quad B(F_2)=\begin{bmatrix}1 & 0 & 0\\ 0 & 0 & 0\\ 0 & 0 & 1\end{bmatrix},\quad B(F_3)=\begin{bmatrix}1 & 0 & 0\\ 0 & 1 & 0\\ 0 & 1 & 0\end{bmatrix}$$

with

$$L=\begin{bmatrix}0 & 0 & 0\\ 0 & 0 & 0\\ 0 & 1 & 0\end{bmatrix}\quad\text{and}\quad U=\begin{bmatrix}0 & 1 & 1\\ 0 & 0 & 0\\ 0 & 0 & 0\end{bmatrix}.$$

This results in

$$B(G)=(I+L+L^2)\,U=B(F_3)\,B(F_2)\,B(F_1).$$

2) Let us reconsider again example 2 above. It turns out that G is

$$\begin{aligned}
g_1(x)&=\bar{x}_3\\
g_2(x)&=\bar{x}_3 x_3=0\\
g_3(x)&=1\\
g_4(x)&=\bar{x}_3+0=\bar{x}_3
\end{aligned}$$

from which

$$B(G)=\begin{bmatrix}0&0&1&0\\0&0&0&0\\0&0&0&0\\0&0&1&0\end{bmatrix}.$$

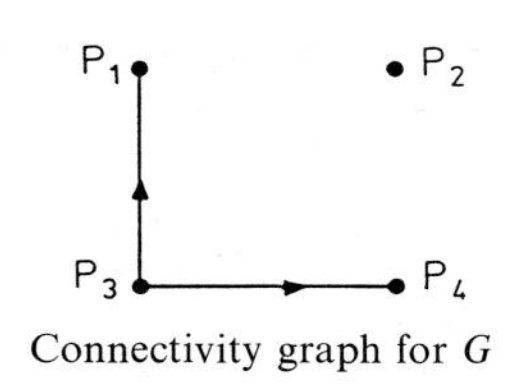

Connectivity graph for G

Since the connectivity graph for G has no circuits it follows that G is contracting. Indeed, the iteration graph for G is simple.

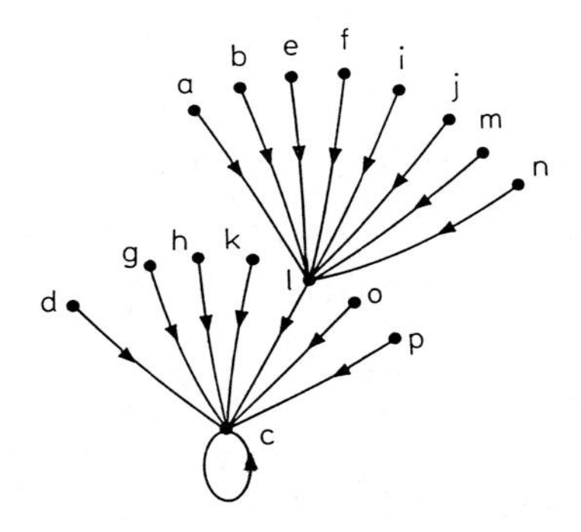

Therefore, for F, one converges to c in no more than *four* steps, whereas for G one converges to c in no more than *two* steps. Here, as well, G is more contracting than F. Below we give a table in which we have recorded the number of steps that separates each element $a, b, \ldots, p$ from c, for both F and G.

	a	b	c	d	e	f	g	h	i	j	k	l	m	n	o	p
for F	4	4	0	1	4	4	2	2	4	4	3	3	4	4	3	3
for G	2	2	0	1	2	2	1	1	2	2	1	1	2	2	1	1

Not only is the maximum number of steps necessary for G for converging (2) *smaller than the corresponding number for F* (4), *but one may also note* *that (in this example*) all elements (but d) "approach" c in the transition from F to G.*

It is therefore more "efficient" to iterate using G, than using F, in order to reach the fixed point c.

6. Comparison of Operating Modes for a Contracting Operator

The result established above that, if F is contracting then G is more contracting, is also valid whenever one compares F with an arbitrary serial-parallel operating mode:

Theorem 4. *If F is contracting relative to d, then all the serial-parallel operators F_τ associated with an arbitrary serial-parallel mode τ of operation are also contracting relative to d.*

More precisely: if F is contracting, then there exists an integer $p \leq n$ such that

$$[B(F)]^p = 0.$$

Also, for that p we have

$$[B(F_\tau)]^p = 0.$$

The iteration graph for F_τ is therefore simple and F_τ is at least as contracting as F.

The proof is formally the same as the proof given for the Gauss-Seidel serial iteration. Here, since F_τ is a block Gauss-Seidel operator associated with F (see Chap. 1, p. 17), the matrix $B(F)$ may be decomposed into $L+U$, where L and U are (apart from a possible permutation of rows and columns) *block* strictly lower triangular and *block* upper triangular (see example p. 37).

One therefore has

$$[B(F)]^p = [L+U]^p = 0$$

from which (see p. 37)

$$[B(F_\tau)]^p \leq [(I+L+\ldots+L^{n-1})\,U]^p = 0$$

using the boolean Stein-Rosenberg theorem. □

The preceding result may moreover be expanded somewhat when one considers the case of a contracting serial-parallel operator F_τ and an operator F_v where the operating mode v is *more sequential* than τ ($\tau \,\alpha\, v$, see Chap. 1, p. 17).

* This is not *always* true, as proven in [79].

Theorem 5. *Let τ and ν be two serial-parallel operating modes satisfying $\tau \alpha \nu$.*

It then follows that if F_τ is contracting relative to d then F_ν is more contracting, or stated differently, then there exists an integer $p \leq n$ such that

$$[B(F_\tau)]^p = 0$$

and such that for the same p

$$[B(F_\nu)]^p = 0.$$

The result follows immediately from the preceding discussion, taking into account Theorem 3 of Chap. 1, p. 19, which shows that F_ν *is in fact a block Gauss-Seidel operator relative to* F_τ. □

Remark. If F is contracting, is there then a *best* serial-parallel process from the point of view of the number of steps necessary to reach the fixed point? The answer is positive, which is shown by:

Theorem 6. *If F is contracting, then there exists an associated (purely serial) process, noted here by H, such that*

$$\forall x \in X \qquad H(x) = \xi.$$

The serial process defined by H is then optimal since it converges to the fixed point ξ in at most one step for an arbitrary starting element x^0.

This is true since if F is contracting then $B(F)$ is a strictly lower triangular matrix (up to a permutation of rows and columns). Assuming therefore, without loss of generality, that $B(F)$ is strictly lower triangular, one verifies that the *Gauss-Seidel process* is optimal. Indeed, independent of the choice of x^0 one has

$$\begin{aligned} x_1^1 &= f_1(\sqcup, \ldots, \sqcup) = \xi_1 \\ x_2^1 &= f_2(\xi_1, \sqcup, \ldots, \sqcup) = \xi_2 \\ &\vdots \\ x_n^1 &= f_n(\xi_1, \ldots, \xi_{n-1}, \sqcup) = \xi_n. \end{aligned}$$

That is to say $x^1 = \xi$ (see also p. 60). □

7. Examples

1) $X = \{0, 1\}^3$ and F is defined by

$$\begin{aligned} f_1(x) &= x_3 \\ f_2(x) &= x_1 + \bar{x}_3 \\ f_3(x) &= 0. \end{aligned}$$

From this we have the incidence matrix of F and its connectivity graph

$$B(F)=\begin{bmatrix}0 & 0 & 1\\ 1 & 0 & 1\\ 0 & 0 & 0\end{bmatrix}.$$

P_2

P_1

P_3

This graph does not have a circuit and F is therefore contracting.

In this simple example, we can calculate all the serial-parallel operators F_τ associated with F and verify that they are all contracting.

After all calculations are done, one is left with four *distinct* serial-parallel operators which we denote by I, II, III, IV. They are given by

I (1, 2, 3); (1, 2) (3); (2) (1, 3); (2) (1) (3);
II (1) (2, 3); (1) (2) (3); (1) (3) (2); (1, 3) (2);
III (3, 2) (1); (2) (3) (1);
IV (3) (1, 2); (3) (1) (2); (3) (2) (1)

The table and the iteration graphs for these four distinct operator modes are given below

	x	I	II	III	IV
a	0 0 0	0 1 0	0 1 0	0 1 0	0 1 0
b	0 0 1	1 0 0	1 1 0	0 0 0	0 1 0
c	0 1 0	0 1 0	0 1 0	0 1 0	0 1 0
d	0 1 1	1 0 0	1 1 0	0 0 0	0 1 0
e	1 0 0	0 1 0	0 1 0	0 1 0	0 1 0
f	1 0 1	1 1 0	1 1 0	0 1 0	0 1 0
g	1 1 0	0 1 0	0 1 0	0 1 0	0 1 0
h	1 1 1	1 1 0	1 1 0	0 1 0	0 1 0

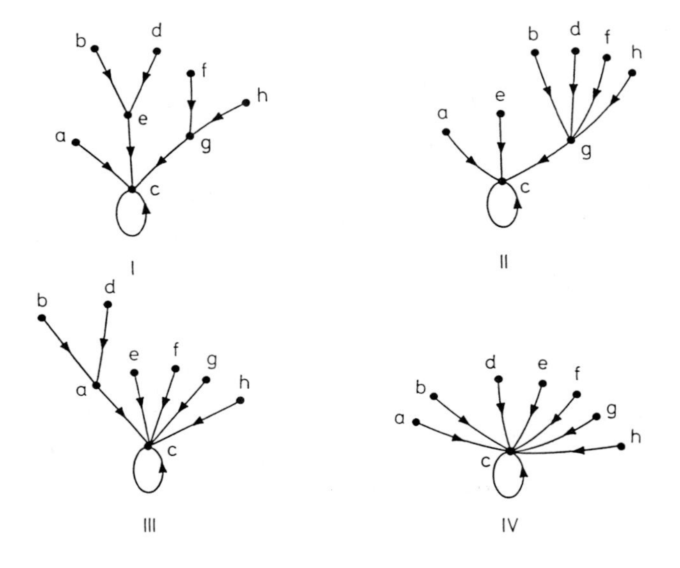

We are furthermore giving the table for the number of iteration steps separating a given element from the fixed point c for each of the 8 elements in $\{0, 1\}^3$ (and for each of the four operator modes I, II, III, IV)

	a	b	c	d	e	f	g	h
for I	1	2	0	2	1	2	1	2
for II	1	2	0	2	1	2	1	2
for III	1	2	0	2	1	1	1	1
for IV	1	1	0	1	1	1	1	1

The various serial-parallel operators are furthermore ordered in the following manner:

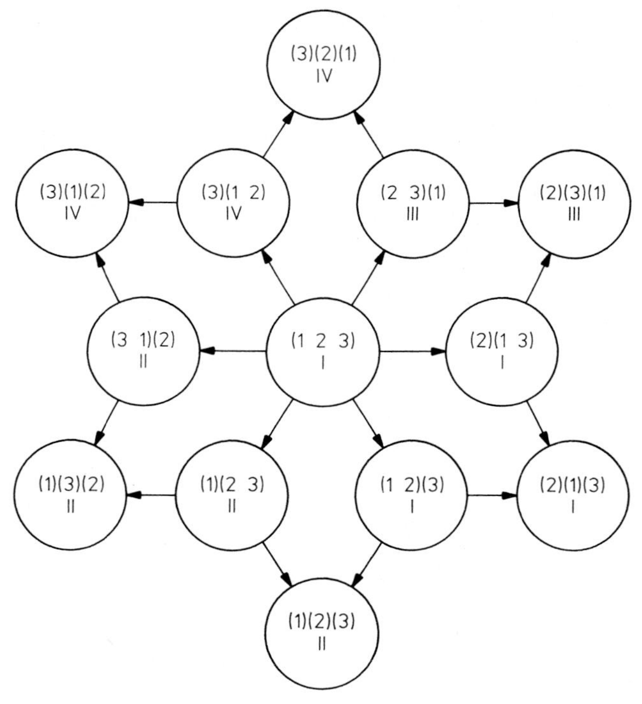

The arrows indicate the transition from an operator to another *more sequential* operator.

This example again provides evidence for the fact that if F is contracting, then the more an associated serial-parallel process is sequential the more contracting it is. In fact, when looking at the above diagram one may work ones way through "the hierarchy of contractions" in the following manner:

$$\mathrm{I} \to \mathrm{II} \to \mathrm{IV} \quad \text{and} \quad \mathrm{I} \to \mathrm{III} \to \mathrm{IV}.$$

Remark. The process IV is *the most contracting* and it corresponds to the serial processes (3) (1) (2) or (3) (2) (1) (see also Theorem 6 above).

Now, with the same incidence matrix as above

$$B(F)=\begin{bmatrix}0 & 0 & 1\\ 1 & 0 & 1\\ 0 & 0 & 0\end{bmatrix}$$

let us change the operator F to

$$f_1(x)=\bar{x}_3$$
$$f_2(x)=x_1 x_3$$
$$f_3(x)=1.$$

For this operator we have $b=(0, 0, 1)$ as the unique fixed point.

After all the calculations are done for this operator we are left with *six* distinct serial-parallel operators which we denote by I, II, III, IV, V and VI. They are given by

I	(1, 2, 3); (2) (1, 3); (1, 2) (3); (2) (1) (3);
II	(1) (2, 3); (1) (2) (3)
III	(3, 2) (1); (2) (3) (1)
IV	(3) (1, 2); (3) (2) (1)
V	(3, 1) (2); (1) (2) (3)
VI	(3) (1) (2).

Below we give the iteration graph for these operators

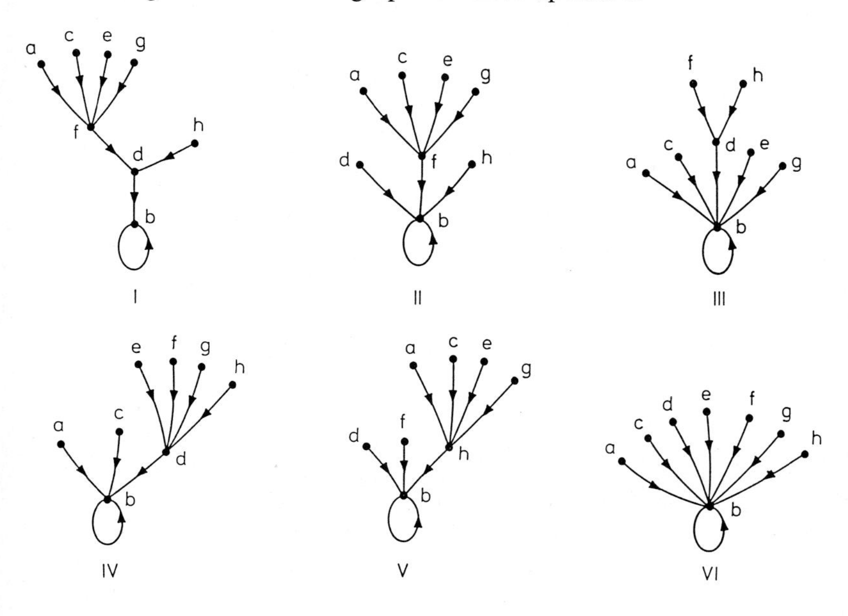

From these the following table is obtained:

	a	b	c	d	e	f	g	h
I	3	0	3	1	3	2	3	2
II	2	0	2	1	2	1	2	1
III	1	0	1	1	1	2	1	1
IV	1	0	1	1	2	2	2	2
V	2	0	2	1	2	1	2	1
VI	1	0	1	1	1	1	1	1

The diagram of the different operating modes is given below (note the analogies with the preceding diagram which is evidently due to the fact that the incidence matrix $B(F)$ is the same for both cases).

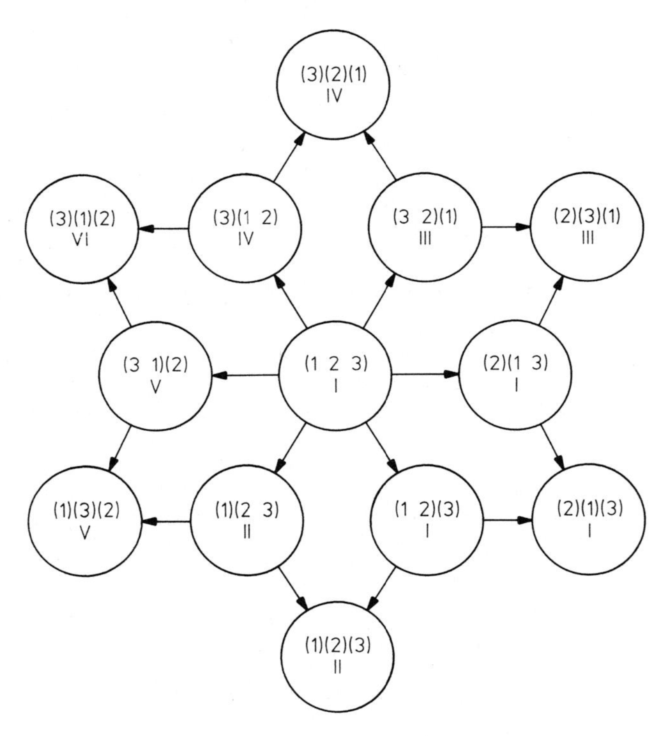

The arrows again indicate the transition from a process to a more sequential (and therefore a more contracting) process. Here, the hierarchy of contractions is

$$\mathrm{I} \to \mathrm{II} \to \mathrm{V} \qquad \mathrm{I} \to \mathrm{III} \to \mathrm{IV}$$

$$\mathrm{I} \to \mathrm{IV} \to \mathrm{VI} \qquad \mathrm{I} \to \mathrm{V} \to \mathrm{VI}.$$

The optimal serial process is (3) (1) (2) (operating mode VI). This optimal order of the variables 3, 1, 2 is clearly determined by the incidence matrix

$$B(F)=\begin{bmatrix} 0 & 0 & 1 \\ 1 & 0 & 1 \\ 0 & 0 & 0 \end{bmatrix}.$$

2) $X=\{0,1\}^4$ and F is defined by

$$\begin{aligned} f_1(x) &= x_3 \\ f_2(x) &= x_1+\bar{x}_3 \\ f_3(x) &= 1 \\ f_4(x) &= x_1\,\bar{x}_2\,x_3. \end{aligned} \qquad \text{(boolean notation)}$$

The incidence matrix and the connectivity graph are then given by

$$B(F)=\begin{bmatrix} 0 & 0 & 1 & 0 \\ 1 & 0 & 1 & 0 \\ 0 & 0 & 0 & 0 \\ 1 & 1 & 1 & 0 \end{bmatrix}.$$

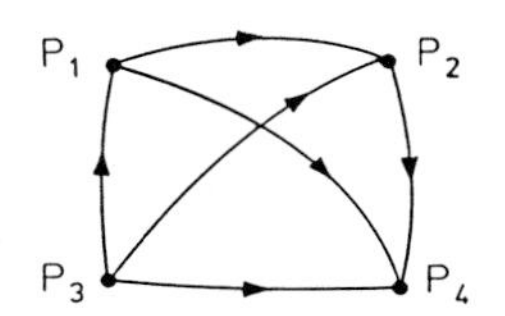

This graph does not have a circuit hence F is contracting.

We consider three operating modes.

The parallel mode (1, 2, 3, 4) defined by the operator F itself.

The serial-parallel mode (1, 2) (3, 4) where we will call the associated operator H.

The serial mode (1) (2) (3) (4) defined by G, the Gauss-Seidel operator.

We get

$$\begin{aligned} h_1(x) &= x_3 & g_1(x) &= x_3 \\ h_2(x) &= x_1+\bar{x}_3 & g_2(x) &= x_3+\bar{x}_3=1 \\ h_3(x) &= 1 & g_3(x) &= 1 \\ h_4(x) &= \bar{x}_1\,x_3 & g_4(x) &= 0. \end{aligned}$$

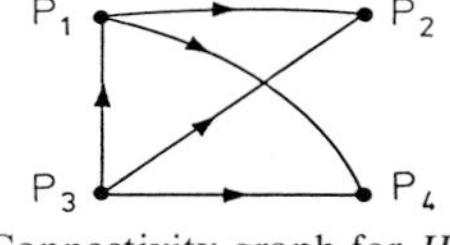

Connectivity graph for H

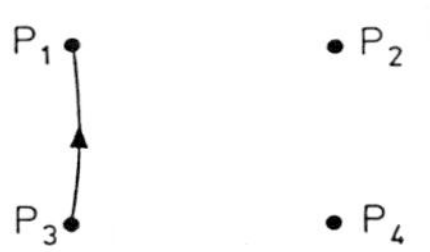

Connectivity graph for G

One verifies that the associated connectivity graphs are without circuits. The corresponding operators H and G are therefore contracting.

After all the calculations have been completed one obtains the iteration graphs

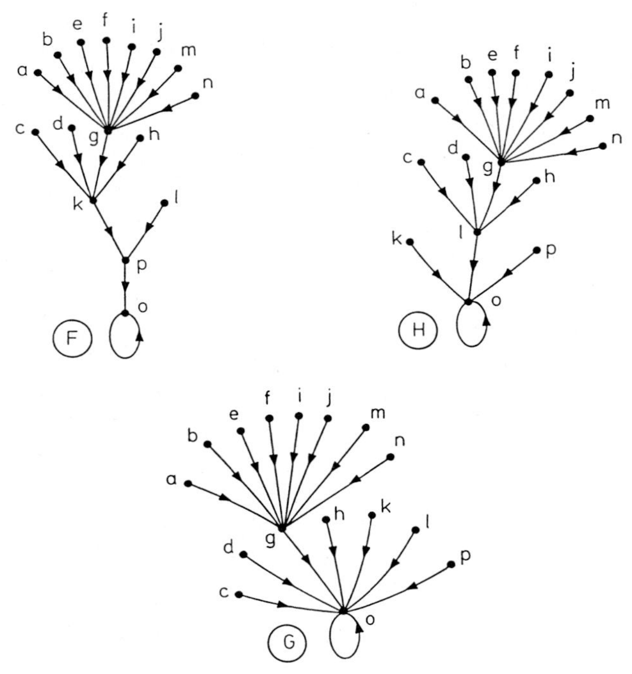

This example clearly shows how one passes from a contracting operator F (process $(1,2,3,4)$) to a more contracting operator H (process $(1,2)(3,4)$) and finally to G (process $(1)(2)(3)(4)$) which is a still more contracting operator.

	a	*b*	*c*	*d*	*e*	*f*	*g*	*h*	*i*	*j*	*k*	*l*	*m*	*n*	*o*	*p*
F	4	4	3	3	4	4	3	3	4	4	2	2	4	4	0	1
H	3	3	2	2	3	3	2	2	3	3	1	1	3	3	0	1
G	2	2	1	1	2	2	1	1	2	2	1	1	2	2	0	1

8. Rounding off: Successive Gauss-Seidelisations

We have seen that (Theorem 3) if F is contracting then the associated Gauss-Seidel operator G is as well, and then at least as contracting as F.

From this one gets the idea that one may iterate this process by forming the Gauss-Seidel operator associated with G and so on. One may show that this process becomes stationary at a limiting operator denoted by $\hat{G}$, which is *triangularized*. More precisely, one has:

Theorem 7. *If F is contracting relative to d, consider the following sequence of operators on X:*

$G^{(0)} = F$

$G^{(1)} = G$ the Gauss-Seidel operator associated with $G^{(0)}$

$\vdots$

$G^{(r)} =$ the Gauss-Seidel operator associated with $G^{(r-1)}$ $\quad (r = 1, 2, \ldots, n)$.

There furthermore exists an index $r \leq n$ such that this process is stationary. In this manner a limit operator $\hat{G}$ is defined that is contracting relative to d, and which has the same fixed point as F. Furthermore $B(\hat{G})$ is strictly upper triangular.

This theorem will not be proven here. The limit fixed point equation is

$$x = \hat{G}(x)$$

and this is equivalent to the starting equation $x = F(x)$.

Since $B(\hat{G})$ is strictly upper triangular it follows that the equation $x = \hat{G}(x)$ may be solved immediately (in the order $\xi_n \xi_{n-1} \ldots \xi_1$ of the components of the fixed point). The starting equation $x = F(x)$ is therefore *triangularized* into an equivalent equation by the successive Gauss-Seidelisations. In fact, this result looses a bit of interest in the discrete context since there always exists a serial-parallel process associated with F (Theorem 6) which already corresponds to a triangular operator.

If we take the preceding example we get

$$\underbrace{\begin{matrix} x_3 \\ x_1 + \bar{x}_3 \\ 1 \\ x_1 \bar{x}_2 x_3 \end{matrix}}_{G^0 = F} \quad \begin{matrix} \\ \\ \rightarrow \\ \\ \end{matrix} \quad \underbrace{\begin{matrix} x_3 \\ x_3 + \bar{x}_3 = 1 \\ 1 \\ 0 \end{matrix}}_{G^{(1)} = G = \hat{G},} \quad \text{from which } \hat{G} = G$$

while with the example (contracting) on p. 64 one gets

$$\underbrace{\begin{matrix} \bar{x}_3 \\ x_1 x_3 \\ 1 \\ x_1 + x_2 x_3 \end{matrix}}_{G^0 = F} \quad \begin{matrix} \\ \rightarrow \\ \\ \\ \end{matrix} \quad \underbrace{\begin{matrix} \bar{x}_3 \\ \bar{x}_3 x_3 = 0 \\ 1 \\ \bar{x}_3 \end{matrix}}_{G^{(1)} = G} \quad \begin{matrix} \\ \rightarrow \\ \\ \\ \end{matrix} \quad \underbrace{\begin{matrix} \bar{x}_3 \\ 0 \\ 1 \\ 0 \end{matrix}}_{G^{(2)} = \hat{G}.}$$

The last operator $\hat{G}$ immediately gives the desired fixed point $\xi = F(\xi) = \hat{G}(\xi)$, it is $\begin{bmatrix} 0 \\ 0 \\ 1 \\ 0 \end{bmatrix} = c$.

9. Conclusions

Starting from the incidence matrix $B(F)$ for an operator F on a product $X = \prod_{i=1}^{n} X_i$ of finite sets, we have defined the notion of a *contraction* for F (relative to the vector distance d). Using this notion we have been able to discuss the existence and uniqueness of a fixed point for F that is always reached in at most n steps independent of the serial-parallel operator-mode associated with F that is employed and independent of the starting element x^0. There exists, moreover, an optimal serial-parallel process that reaches the fixed point in one step.

Clearly, this notion of contraction is rather strong. In practical cases it is rarely realized due to the presence of several basins in the iteration graph of F, having fixed point(s) or cycle(s), the importance of the starting configuration and the operator mode used and so on.

Nevertheless, this study is a tentative step in the direction of providing a mathematical foundation for the description of the behaviour of an automata network. In effect, what might be interesting is to see how the ideas and results of numerical analysis in the context of iterative methods for systems of equations (Refs. [15] to [46]) may be transposed into the context of discrete iterations. This transfer is made possible by the use of the metric tool d as well as the boolean versions of the Perron-Frobenius and Stein-Rosenberg theorems. The information carried by the incidence matrix $B(F)$ is determining. This information is of the form all or nothing, either f_i depends on x_j or it does not. The information does not describe the *nature* of the dependency if it exists. The notion of a *discrete derivative* (Chap. 6) will provide more detail for this dependency.

We also note here that the study we have executed on the *synchronous* (serial-parallel) processes might also be extended (in the discrete metric context) to *asynchronous* iterations (see [15], [23] and [166]).

5. Comparison of Operating Modes

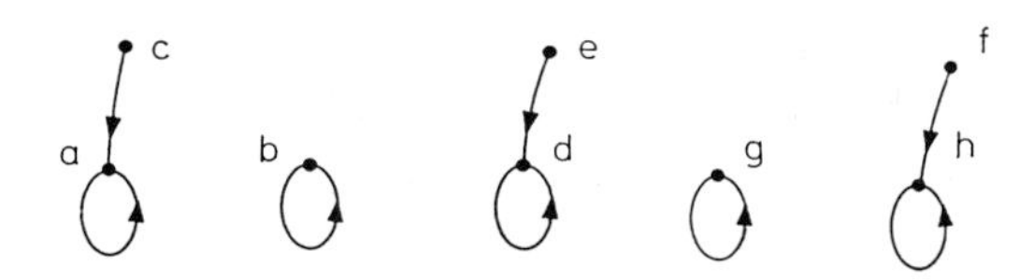

In the preceding chapter we were able to establish results on the behaviour of discrete iterations using the vector distance d as well as the (rather strong) notion of boolean contraction. These results were essentially transplanted from the context of continuous iterations into the discrete context. Under the assumption of contraction we were also able to make some comparisons between different operating modes ("More sequential implies more contracting").

In this chapter *we further compare the operating modes, but now outside the notion of contraction.*

Let us recall (see Chap. 1) that for a given F, the various serial-parallel operating modes that one may define on the associated automata network appears as methods of successive approximations on various operators obtained formally from F by substitution of variables. These various operating modes are then characterized by their iteration graphs.

If we now look at the examples where we were able to describe the graphs completely, then it seems that these graphs often have a "familiar feeling". Indeed, for each example the iteration graph always have the same fixed points (stable configurations of cells when interpreted in terms of automata networks). In some of the examples one can actually see that there are other frequent (but not systematic) phenomena. These occur when passing from a given mode of operation to a *more sequential* mode of operation. The following phenomena are observed (see § 2 and § 4, examples and counterexamples):

– *Bursting* of connected components (basins) without fixed points to the advantage of basins containing a fixed point.

– *Aggregation* of the latter due to the absorption of new points without losing any points that are already there.

– *Implosion* of basins containing a fixed point. For a given starting element, when iterating towards a fixed point in a given operating mode, *the same fixed point is reached at least as rapidly* in a more sequential operating mode.

For a given F, we will now attempt to formulate assumptions that will enable us to *predict* the occurrence of the above phenomena. The assumptions are of a metric kind, utilizing the same metric tool as in the preceding chapter (the boolean vector distance d). To this tool we add the key notion of *monotonicity*. (Contraction and monotonicity are in fact the basic notions used in the study of iterations in a continuous setting.)

So, once more, we try to transpose the metric results from the continuous setting into the discrete framework.

1. Comparison of Serial and Parallel Operating Modes

Let us recall the setting. $X = \prod_{i=1}^{n} X_i$ is a cartesian product of n finite sets X_i, and F is a map of X into itself. Furthermore, G denotes the associated Gauss-Seidel operator for F.

We are interested in studying the iteration graph for G starting from the iteration graph for F. We already know that both these graphs have the same fixed points (stable configurations). Here we will therefore attempt to say some more about the connections between these graphs.

Let us start by examining two elementary examples.

1) $X = \{0, 1\}^3$, F is defined by its iteration graph

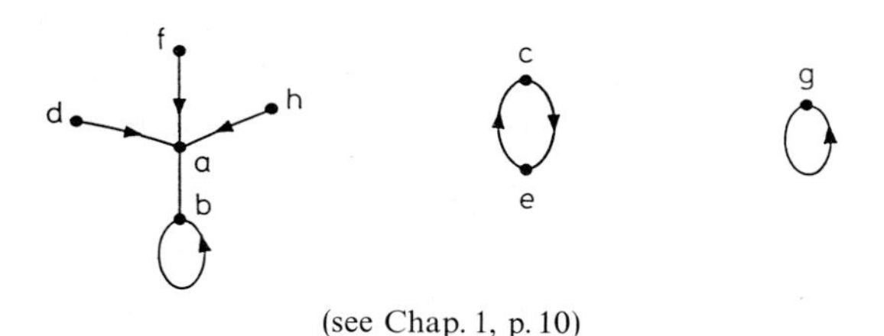

(see Chap. 1, p. 10)

The iteration graph for G is then the following:

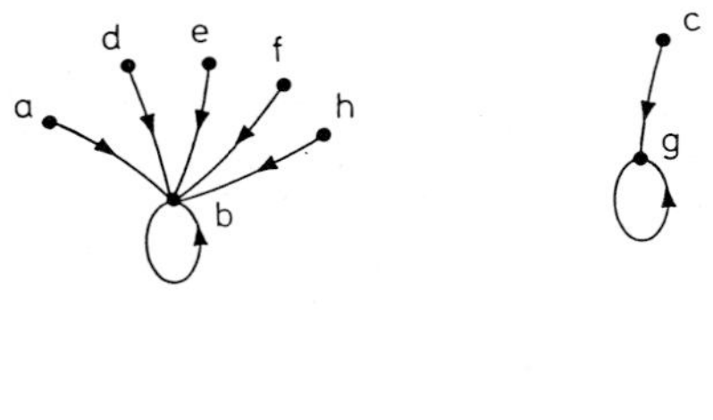

Clearly, in this very simple example, the serial iteration (*G*) *is algorithmically better than the parallel iteration* (*F*).

When passing from F to G, the basin that had a cycle burst. The other phenomena, indicated above, of aggregation and implosion, are happening to the other basins. *A fixed point is therefore reached more frequently as well as more rapidly when iterating with G as compared to F.*

2) Consider now the following example:

$X = \{0, 1\}^3$, F is defined by its iteration graph

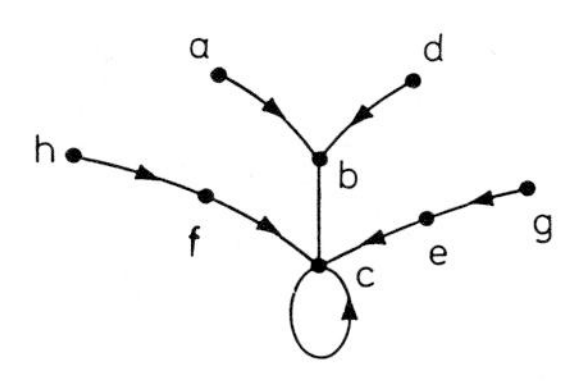

The following is the iteration graph of the associated Gauss-Seidel operator:

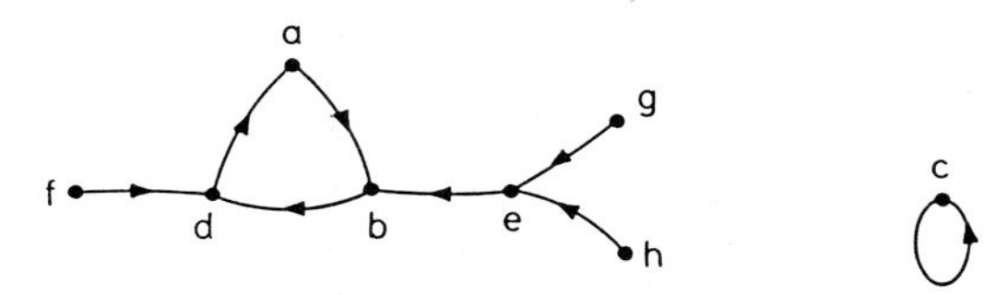

In this example, whereas the iteration graph for F is simple, the iteration graph for G is not. In fact, the fixed point c has been isolated and the rest of the graph has been turned into a basin having a cycle.

It is clear in this case that, from the point of view of searching for a fixed point, the serial process is a disaster, whereas the parallel process always delivers the fixed point in no more than two steps. (The two examples above may be made more elaborate by extending them to $\{0, 1\}^n$. See also example 2) pp. 13 and 14.)

In what follows, we will establish conditions on F assuring that G is "*algorithmically better*" (for searching for a fixed point) than F.

ξ denotes a fixed point of F (hence of G) whose existence is a priori assumed.

The notation $x_{\tilde{F}}\, \xi$ means that x belongs to the basin containing ξ in the iteration graph for F (this fits well with the equivalence relation defined on p. 2. Indeed, to say that $x_{\tilde{F}}\, \xi$ is the same as to say that there exists an integer $p \geq 0$ such that $F^p(x) = \xi$).

In the same manner, all points y belonging to the basin containing ξ in the iteration graph for G are noted by $y_{\tilde{G}}\, \xi$.

Definition. The serial iteration is said to be *better* than the parallel iteration if the following property (P) is true:

For all fixed points ξ of F (and G) and for all $x_{\hat{F}}\,\xi$ (there exists then a $p \geq 0$ such that $F^p(x) = \xi$) one has

$$G^p(x) = \xi \qquad (P)$$

(from which clearly $x_{\hat{G}}\,\xi$).

This property expresses the fact that, in the passage from the iteration graph for F to the one for G:

– *the size of each basin having a fixed point is at least maintained* (and it will possibly grow, to the detriment of basins having a cycle). In other words, if ξ is a fixed point of F (and G) and if $x_{\hat{F}}\,\xi$ one also has $x_{\hat{G}}\,\xi$.

– *furthermore, each basin having a fixed point can only implode*, in the sense that if ξ is a fixed point of F (and G) and if p iteration steps are required to reach ξ for some x when iterating with F, *then at most p iteration steps are required for that same x when iterating with G.*

Property (P) expresses the fact that in the passage from a parallel iteration to a serial iteration

– the convergence phenomena are amplified

– the phenomena of cycling are attenuated.

The example 1) above illustrates the property (P) whereas example 2) does not.

Definitions. If ξ is a fixed point of F, then F is called *ξ-monotone* if the inequality

$$d(x, \xi) \leq d(y, \xi) \qquad \text{(where } x \text{ and } y \in X)$$

implies the inequality

$$d(F(x), \xi) \leq d(F(y), \xi)$$

from which, clearly for all r

$$d(F^r(x), \xi) \leq d(F^r(y), \xi).$$

F is called *ξ-non-expansive* if

$$\forall x_{\hat{F}}\,\xi \quad \text{one has} \quad d(F(x), \xi) \leq d(x, \xi)$$

from which clearly for all r

$$d(F^r(x), \xi) \leq d(x, \xi).$$

Finally, F and G are called *ξ-ordered* if

$$\forall x_{\hat{F}}\,\xi \quad \text{one has} \quad d(G(x), \xi) \leq d(F(x), \xi).$$

Finally, recall (see Chap. 1, p. 12) that for all $x=(x_1, \ldots, x_n)$ in X, we write

$$F_i(x)=\begin{bmatrix} x_1 \\ \vdots \\ f_i(x_1, \ldots, x_n) \\ \vdots \\ x_n \end{bmatrix}$$

and we have

$$G=F_n \circ \ldots \circ F_2 \circ F_1.$$

Theorem 1. *Let ξ be a fixed point of F. Now, if F is ξ-monotone, then G is as well. Furthermore, for all x satisfying $F(x)=\xi$, one has $G(x)=\xi$.*

The following two points are easy to verify:

– F is ξ-monotone if and only if F_i is $(i=1, 2, \ldots, n)$ (ξ is clearly a fixed point for F_i, $i=1, 2, \ldots, n$).

– The product of two ξ-monotone operators is ξ-monotone.

Then F is ξ-monotone if and only if the F_i's are, from which we get the result that $G=F_n \circ \ldots \circ F_1$ is ξ-monotone.

Now, let x be such that $F(x)=\xi$, that is to say we have

$$f_i(x_1, \ldots, x_n)=\xi \qquad (i=1, 2, \ldots, n).$$

We now show by induction that $G(x)=\xi$, that is to say

$$g_i(x_1, \ldots, x_n)=\xi_i \qquad (i=1, 2, \ldots, n).$$

This is clearly true for $i=1$ since $g_1=f_1$.

Assume, therefore, for the induction, that

$$g_j(x_1, \ldots, x_n)=\xi_j \qquad (j=1, 2, \ldots, i-1).$$

Then consider

$$\begin{aligned} g_i(x) &= f_i(g_1(x), \ldots, g_{i-1}(x), x_i, \ldots, x_n) \\ &= f_i(\underbrace{\xi_1, \ldots, \xi_{i-1}, x_i, \ldots, x_n}_{\hat{x}}). \end{aligned}$$

Clearly $d(\hat{x}, \xi) \leq d(x, \xi)$ and F is ξ-monotone. Then

$$0 \leq d(F(\hat{x}), \xi) \leq d(F(x), \xi)=d(\xi, \xi)=0$$

which implies that

$$F(\hat{x})=\xi.$$

The i-th component of this relation may be written

$$f_i(\xi_1, \ldots, \xi_{i-1}, x_i, \ldots, x_n)=\xi_i$$

from which we finally have

$$g_i(x) = f_i(\hat{x}) = \xi_i.$$

The relation $G(x) = \xi$ is therefore proven by induction. □

Theorem 2. *Let ξ be a fixed point of F. If F is ξ-monotone and ξ-non-expansive, then F and G are ξ-ordered (the converse is not true).*

With the above assumptions let $x_{\tilde{F}}\,\xi$. We have to show that

$$d(G(x), \xi) \leq d(F(x), \xi).$$

The first component of this inequality is true since

$$\delta_1(g_1(x), \xi_1) = \delta_1(f_1(x), \xi_1).$$

Assume therefore the induction hypothesis that

$$\delta_j(g_j(x), \xi_j) \leq \delta_j(f_j(x), \xi_j) \qquad j = 1, 2, \ldots, i-1$$

and let us show that

$$\delta_i(g_i(x), \xi_i) \leq \delta_i(f_i(x), \xi_i).$$

We have that

$$\delta_i(g_i(x), \xi_i) = \delta_i(f_i(\underbrace{g_1(x), \ldots, g_{i-1}(x), x_i, \ldots, x_n}_{u}), \xi_i)$$

$$= \delta_i(f_i(u), \xi_i).$$

If we let $v = (f_1(x), \ldots, f_{i-1}(x), x_i, \ldots, x_n)$ then the recurrence assumption implies that

$$d(u, \xi) \leq d(v, \xi).$$

Moreover, $x_{\tilde{F}}\,\xi$ and F is ξ-non-expansive. It results in

$$d(v, \xi) \leq d(x, \xi).$$

This means, finally, that

$$d(u, \xi) \leq d(x, \xi).$$

Since F is ξ-monotone, this furthermore implies that

$$d(F(u), \xi) \leq d(F(x), \xi)$$

from which (i-th component of this inequality)

$$\delta_i(g_i(x), \xi_i) \leq \delta_i(f_i(x), \xi_i).$$

One has therefore shown by induction that F and G are ξ-ordered. The converse of this theorem is false, see Example 1, p. 88. □

Theorem 3. *Let ξ be a fixed point of F. Furthermore, let F be ξ-monotone and F and G be ξ-ordered.*

Assume also that $x_{\bar{F}}\,\xi$ *(x belongs to the basin defined by* ξ *in the iteration graph for F).*

Then for all integers $r \geq 0$ *one has*

$$d(G^r(x), \xi) \leq d(F^r(x), \xi).$$

This property is certainly true for $r=0$ ($F^0 = G^0 = I$) as well as for $r=1$ (F and G are ξ-ordered). Assume therefore that it is valid for $r=2, 3, \ldots, i-1$. We then have for all $x_{\bar{F}}\,\xi$ that

$$d(\underbrace{G^{i-1}(x)}_{y}, \xi) \leq d(\underbrace{F^{i-1}(x)}_{z}, \xi). \tag{A}$$

Since $x_{\bar{F}}\,\xi$ and since z is a descendant of x in the iteration on F it also follows that $z_{\bar{F}}\,\xi$. Since F and G are ξ-ordered it also follows that

$$d(G(z), \xi) \leq d(F(z), \xi). \tag{B}$$

On the other hand, since F is ξ-monotone then so is G (Theorem 1). From the inequality (A) it therefore follows that

$$d(G(y), \xi) \leq d(G(z), \xi). \tag{C}$$

From (C) and (B) we conclude that

$$d(G(y), \xi) \leq d(F(z), \xi)$$

which from the definition of y and z results in

$$d(G^i(x), \xi) \leq d(F^i(x), \xi)$$

and the desired inequality is proven by induction. □

Remark. We may moreover show under the assumptions of the previous theorem that we have

$$\forall x_{\bar{F}}\,\xi \quad \text{and for all} \quad r \geq 0$$

$$d(G^r(x), \xi) \leq d(H_r \ldots H_2 H_1(x), \xi) \leq d(F^r(x), \xi)$$

where H_i may be either F or G ($i = 1, 2, \ldots, r$).

Theorem 4. *Let* ξ *be a fixed point of F. Furthermore let F be* ξ*-monotone and F and G* ξ*-ordered.*

Assume also that $x_{\bar{F}}\,\xi$ *such that there exists an integer* $p \geq 0$ *for which* $F^p(x) = \xi$.

Then there exists an integer $q \leq p$ *such that* $G^q(x) = \xi$ *or equivalently* $x_{\bar{G}}\,\xi$.

It suffices to show that $G^p(x) = \xi$. This is, however, an immediate consequence of the previous theorem since

$$0 \leq d(G^p(x), \xi) \leq d(F^p(x), \xi) = 0.$$ □

We now wish to give a geometric interpretation for Theorem 4. For this purpose let ξ be a fixed point for F and let x belong to the basin defined by ξ in the iteration graph for F. A consequence of this is of course that there exists a p such that $F^p(x)=\xi$.

Under the given assumptions on x one is also assured that x is a member of the basin defined by ξ in the iteration graph for G (ξ is also clearly a fixed point of G). Better: When iterating with G, starting from x (serial iteration) one reaches ξ in at most as many steps as are required to reach ξ from x when iterating with F (parallel iteration).

We have therefore clearly shown that when passing from F to G, the basin defined by ξ

- conserves at least its effective size (aggregation)
- might implode

provided F is ξ-monotone and F and G are ξ-ordered. In order that the second condition holds true it suffices, according to Theorem 2, that F is ξ-non-expansive. We now recapitulate the above discussion:

$$\left.\begin{array}{l} F \quad \xi\text{-monotone} \Rightarrow G \ \xi\text{-monotone} \\ \text{Moreover if } F(x)=\xi \text{ then } G(x)=\xi \end{array}\right\} \quad \text{(Theorem 1)}$$

$$\left.\begin{array}{l} F \quad \xi\text{-monotone} \\ F \quad \xi\text{-non-expansive} \end{array}\right\} \Rightarrow F \text{ and } G \ \xi\text{-ordered} \quad \text{(Theorem 2)}$$

$$\left.\begin{array}{l} F \quad \xi\text{-monotone} \\ F \text{ and } G \ \xi\text{-ordered} \end{array}\right\} \Rightarrow \begin{array}{l} \text{aggregation and implosion of the basin defined by} \\ \xi \text{ when going from } F \text{ to } G. \quad \text{(Theorem 4)} \end{array}$$

By assuming conditions on *all* the fixed points of F we clearly have the following theorem:

Theorem 5. *If F is ξ-monotone and if F and G are ξ-ordered for all the fixed points ξ of F (and G) then the serial iteration is better than the parallel iteration.*

(Property (P) is assured.)

Indeed, in passing from F to G one has the aggregation and implosion of each basin defined by a fixed point to the detriment of the basins without a fixed point.

It should moreover be noted that if the iteration graph for F only has basins with fixed points then under the assumptions for the above theorem, it follows that each basin keeps its own elements and clearly also gains no other elements when passing to the serial iteration defined by G (the basins may, however, implode).

The following examples are given to illustrate the above points:

$$F(x)=x \qquad \forall x \in X$$
$$F(x)=\text{const.} \qquad \forall x \in X.$$

In these two cases F is ξ-monotone and ξ-non-expansive for all fixed points ξ and furthermore $G=F$.

Remarks. (1) We might consider the following more restrictive conditions:
- *monotonicity* of F:

$$d(x, z) \leq d(y, z) \Rightarrow d(F(x), F(z)) \leq d(F(y), F(z))$$

which clearly implies that F is ξ-monotone at each fixed point ξ of F.
- *non-expansion* of F:

$$\forall x, y \in X \qquad d(F(x), F(y)) \leq d(x, y)$$

which clearly implies that F is ξ-non-expansive at each fixed point ξ of F.
- *F and G are ordered* if

$$\forall x, y \in X \qquad d(G(x), G(y)) \leq d(F(x), F(y))$$

which clearly implies that F and G are ξ-ordered at each fixed point ξ of F (and G).

Then from the preceding it is clear that if F is monotone, with F and G ordered, then the serial-iteration is *better* than the parallel iteration.

It is moreover possible to establish results analogous to the preceding theorems directly using the above notions. The proofs are quite similar.

The notions that we finally retained (ξ-monotonicity, ξ-non-expansion, ξ-ordering) are less restrictive since they are more local.

(2) Let ξ be a fixed point of F and x and y two elements of X. If the inequality

$$d(x, \xi) \leq d(y, \xi)$$

holds, then we write

$$x \propto y.$$

The relation $\propto$ is clearly

reflexive and *transitive* on X: it is a *preorder*.

Then the notion of *ξ-monotonicity for F* may be written as

$$x \propto y \Rightarrow F(x) \propto F(y).$$

This is the monotonicity of F relative to the preorder $\propto$. The notion of *ξ-non-expansion of F* may be written as

$$\forall x_{\bar{F}}\, \xi \qquad F(x) \propto x.$$

The notion of *ξ-ordering* may be written as

$$\forall x_{\bar{F}}\, \xi \qquad G(x) \propto F(x).$$

Clearly the relation [$x \propto y$ *and* $y \propto x$] (which is written $d(x, \xi) = d(y, \xi)$) is an equivalence relation on X. Therefore $\propto$ introduces an order on the quotient space.

Moreover, if $X=\{0,1\}^n$ then $d(x,\xi)=d(y,\xi)$ implies that $x=y$. $\propto$ is therefore an *order* relation on X.

With this remark one is led to the study of a monotone operator in the usual sense.

2. Examples

1) Let us reconsider the preceding example 1)

$X=\{0,1\}^3$ F is defined by

$f_1(x)=x_2\bar{x}_3$

$f_2(x)=x_1\bar{x}_3$

$f_3(x)=\bar{x}_1\bar{x}_2$ (see also p. 10).

The iteration graphs for F and G are

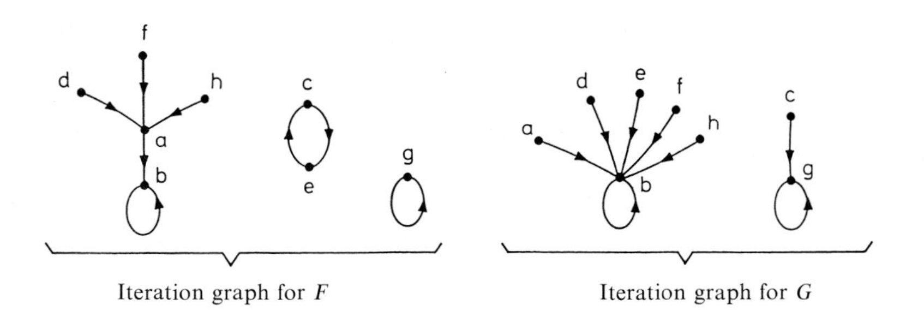

Iteration graph for F Iteration graph for G

One may verify in this example, *after some laborious calculations*, that:

F is b-monotone and g-monotone.

F and G are b-ordered (and clearly g-ordered).

F is *not* b-non-expansive.

F is clearly g-non-expansive.

This example illustrates the preceding results. In the transition from F to G one observes the bursting of the basin with a cycle to the gain of the other basins. Furthermore, one observes the implosion of the basin defined by b. We also remark that F and G are b-ordered even though F, which is b-monotone, is *not* b-non-expansive (cf. Theorem 2).

2) Let us consider the following network of 3 "neurons" (see Chap. 1, p. 10):

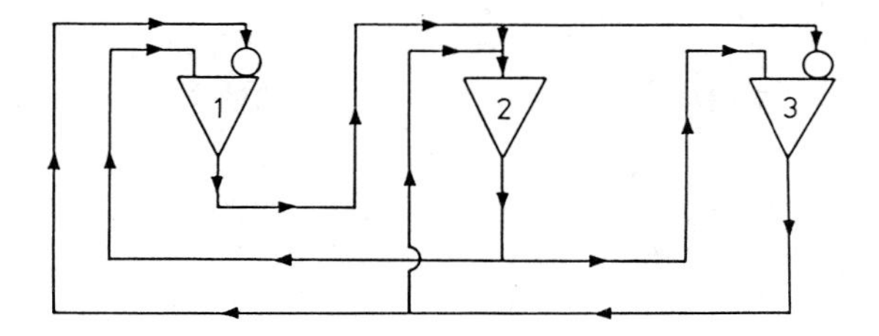

In this example with 3 boolean variables, one has

$$f_1(x_1, x_2, x_3) = x_2 \bar{x}_3$$
$$f_2(x_1, x_2, x_3) = x_1 + x_3$$
$$f_3(x_1, x_2, x_3) = \bar{x}_1 x_2.$$

From this one obtains the tables for F and G as well as their iteration graphs

x	$F(x)$	$G(x)$
0 0 0	0 0 0	0 0 0
0 0 1	0 1 0	0 1 1
0 1 0	1 0 1	1 1 0
0 1 1	0 1 1	0 1 1
1 0 0	0 1 0	0 0 0
1 0 1	0 1 0	1 1 0
1 1 0	1 1 0	1 1 0
1 1 1	0 1 0	0 1 1

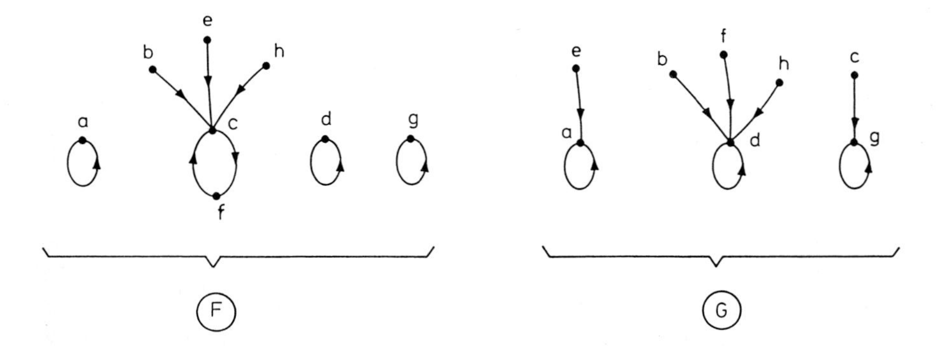

It is clear that G (serial iteration) is *better* than F (parallel iteration). Now, F is *not* ξ-monotone, in particular for $\xi = a$. Thus the conditions stated in Theorem 5 are only *sufficient* conditions.

3. Extension to the Comparison of Two Serial-Parallel Modes of Operation

In the previous section we compared the serial and the purely parallel modes of operation under certain given assumptions. This comparison is now generalized, without difficulty, to the case when two serial-parallel modes of operation are compared.

Let τ and v therefore be two serial-parallel modes of operation such that v is *more sequential* than τ (that is, $\tau \propto v$, see also Chap. 1, p. 17). *We now recall that if F_τ and F_v are the associated operators, then F_v is a block Gauss-Seidel operator relative to F_τ* (Chap. 1, Theorem 3, p. 19).

If ξ is now a fixed point for F (and hence also for F_τ and F_v) then the following theorems are proven formally in the same manner as Theorems 1, 2, 3, 4 and 5.

Theorem 1'. *If F_τ is ξ-monotone, then so is F_v. Furthermore, for all x such that $F_\tau(x)=\xi$ one also has $F_v(x)=\xi$.*

Theorem 2'. *If F_τ is ξ-monotone and ξ-non-expansive then F_τ and F_v are ξ-ordered (and the converse is false).*

Theorem 3'. *Let F_τ be ξ-monotone and let F_τ and F_v be ξ-ordered. Furthermore, let $x_{\tilde{F}_\tau}\xi$ (x belongs to the basin defined by the fixed point ξ in the iteration graph for F_τ).*

Then for every integer $r \geq 0$ one has

$$d(F_v^r(x), \xi) \leq d(F_\tau^r(x), \xi).$$

Theorem 4'. *Assume again that F_τ is ξ-monotone and that F_τ and F_v are ξ-ordered.*

Let $x_{\tilde{F}_\tau}\xi$, such that there exists an integer $p \geq 0$ for which

$$F_\tau^p(x)=\xi.$$

Then there exists an integer $q \leq p$ such that

$$F_v^q(x)=\xi$$

which implies that $x_{\tilde{F}_v}\xi$.

The last theorem allows us to compare the iteration graphs of F_τ and F_v under the given assumptions. To be more precise: all the points belonging to the basin defined by ξ in the iteration graph for F_τ are necessarily found in the basin defined by ξ in the graph for F_v. Thus, in passing from τ to v, that is, from one serial-parallel process to a more sequential serial-parallel process, the basin defined by ξ can only *aggregate*. Furthermore this basin will *implode* in the sense that if x reaches ξ in the operator mode τ in p iteration steps then in the operator mode v it reaches ξ in *at most* p steps.

Remark. The purely parallel process is defined by $a = (1, 2, \ldots, n)$. This means that $F = F_a$. For all other serial-parallel processes τ one clearly has $a \alpha \tau$.

Consequently, if F is ξ-monotone for a certain fixed point ξ then so is F_τ for an arbitrary τ (Theorem 1′). If we now have a third serial-parallel process v satisfying $\tau \alpha v$ then it suffices that F_τ and F_v are ξ-ordered in order that the phenomena studied will occur when passing from F_τ to F_v (Theorem 4′ above). For that, it is moreover sufficient that F_τ is ξ-non-expansive. In this case, it is clear that F_τ and F_v are ξ-ordered as long as τ and v satisfy $\tau \alpha v$ according to Theorem 2′.

In particular, if $F = F_a$ is both ξ-monotone and ξ-non-expansive then F and F_v are ξ-ordered for all serial-parallel processes v. This means that the phenomena reported in Theorem 4′ will necessarily occur between F and F_v for an arbitrary v. For a counter-example see the example below. There F is ξ-monotone for each of the two fixed points b and g. F is, however, surely not ξ-non-expansive (since the phenomena studied between F and F_v are not necessarily occurring for all v, for example for $v = (3)\ (1\ 2)$). A direct verification confirms this result.

Again imposing conditions on *all* the fixed points of F one obtains the generalization of Theorem 5:

Theorem 5′.

If F_τ is ξ-monotone and if F_τ and F_v are ξ-ordered } *for all fixed points ξ of F*

(and hence also of F_τ and F_v) then the serial-parallel procedure v is better than the procedure τ.

4. Examples

Let us consider Example 1) above once more:

The following diagram shows the graphs of *all* the serial-parallel iterations associated with F.

Arrows are used to indicate the passing from one operator mode to a more sequential operator mode. A double arrow indicates the occurrence of one or more of the phenomena of bursting, aggregation and implosion. It should be noted that whenever a basin with a cycle bursts then one also has aggregation and implosion of basins with fixed points.

We notice that for the basin defined by the fixed point b the phenomena of Theorem 4′ are produced when passing from F to $G = (1)(2)(3)$ but not when passing from F to an arbitrary serial-parallel mode.

F is here b-monotone but *not* b-non-expansive.

On the other hand, for the basin defined by the fixed point g, the phenomena of Theorem 4′ are produced when passing from F to *any* of the serial-parallel processes. This illustrates the remark to Theorem 4′. It is noted that F is g-monotone and also g-non-expansive. (Then F and F_v are all g-ordered for any v.)

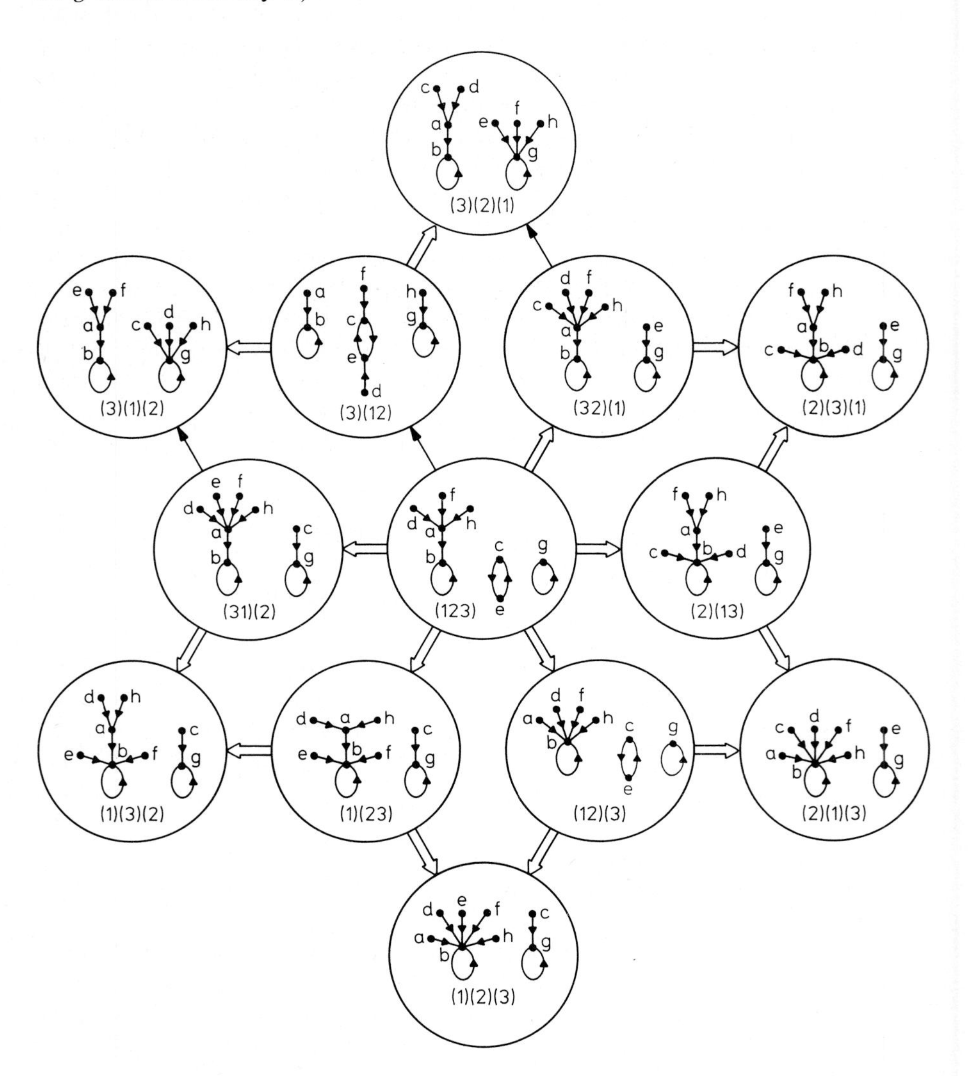

5. Conclusions

We have used some relatively simple notions (ξ-monotonicity, ξ-non-expansion and ξ-ordering) in the context of discrete iterations. Based on these notions we were able to make some comparisons of the iteration graph of a serial-parallel procedure with the iteration graph of a more sequential procedure. It turns out that the latter is often better, although this is certainly not always the case. In fact it might even be that (see example p. 81) the parallel process always converges to a fixed point whereas the serial process has only a cycle and an isolated fixed point.

Admittedly some laborious calculations were required in order to verify the stated conditions for a given example. Moreover these conditions are only *sufficient* conditions for the phenomena under investigation.

Stated briefly then, it is clear that there remain more things to discover and to investigate in further depth in this area.

Having stated this, it is also clear that the algebraic formalization and the topological tool that we have introduced indeed are of interest since they do permit us to prove statements about discrete iterations and their graphs, or stated otherwise, to introduce a mathematical structure for the subject.

6. The Discrete Derivative and Local Convergence

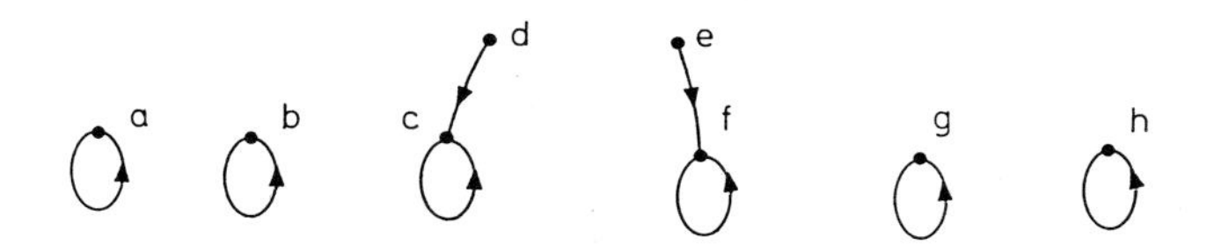

In Chap. 2 we introduced a metric tool which was then used to develop the concepts of contraction and monotonicity. With the aid of these concepts we were then able to study certain behaviours of discrete iterations. The incidence matrix $B(F)$ for an operator F was furthermore found to play a basic role within the framework of contractions through the information it carried (f_i depends or does not depend on x_j). The convergence results obtained were of a *global* kind ("independent of the initial configuration, the iteration converges to a fixed point ..."). These results transpose the classical results from the continuous context into the discrete framework (see for example [32]).

In this chapter we work in a "finer" manner (less globally) and we study the *local behaviour* of discrete iterations (on $\{0,1\}^n$).

We introduce the basic notion of the *discrete derivative* of F for all x. This derivative is the analog, in the discrete setting, of the Jacobian of F in the continuous setting. In this manner we are able to transpose results of classical analysis into the discrete context in a meaningful manner.

1. The Discrete Derivative

We have the following setting. X is now the set $\{0,1\}^n$, the set of vectors having n components either 0 or 1

$$x \in \{0,1\}^n \qquad x=(x_1, \ldots, x_n) \qquad x_i \in \{0,1\}.$$

Among the 2^n such vectors we distinguish the n "basis vectors" e_i whose components are all 0 except for the i-th which is 1 ($i=1,2,\ldots,n$).

Clearly we may identify $\{0,1\}^n$ with the vertices of the n-cube. In this interpretation one notes for $x=(x_1,\ldots,x_n)$ that

$$\tilde{x}^j=(x_1,\ldots,\bar{x}_j,\ldots,x_n)$$

is the j-th *neighbour* of x ($j=1,2,\ldots,n$).

The *immediate neighbourhood* of x is defined to be the set V_x of vertices of the n-cube formed by x and its n neighbours

$$V_x=\{x,\tilde{x}^1\ldots\tilde{x}^n\}$$

$n=3$

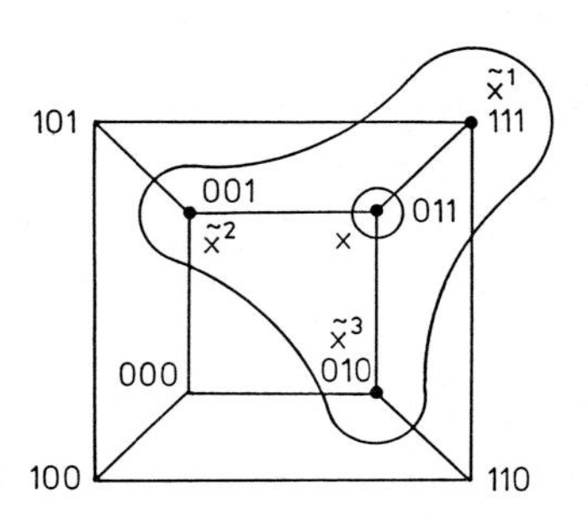

The immediate neighbourhood of $x=(0\ 1\ 1)$ is

$$V_x=\{(0\ 1\ 1),\ (1\ 1\ 1),\ (0\ 0\ 1),\ (0\ 1\ 0)\}.$$

Clearly, if $y\in V_x$ then $x\in V_y$.

We use the usual boolean vector distance d on $\{0,1\}^n$, which is defined as follows. Let $x=(x_1,\ldots,x_n)$ and $y=(y_1,\ldots,y_n)$ both be in $\{0,1\}^n$. Then

$$d(x,y)=(\delta_i(x_i,y_i))$$

where

$$\delta_i(x_i,y_i)=1 \quad \text{if } x_i\neq y_i$$
$$=0 \quad \text{if } x_i=y_i.$$

We recall that d satisfies the following axioms:

$$\begin{cases} d(x,y)=d(y,x) \\ d(x,y)=0\Rightarrow x=y \\ d(x,z)\leq d(x,y)+d(y,z) \end{cases}$$

↑ componentwise inequality in $\{0,1\}^n$ $(0\leq 0\leq 1\leq 1)$

↑ boolean sum in $\{0,1\}^n$ $(1+1=1)$

Remember that even though d takes its values in $\{0,1\}^n$ and not in $\mathbb{R}^n$, d is still a metric tool that is topologically equivalent to the discrete metric on

$\{0,1\}^n$. We also have that the convergent sequences are the sequences that are stationary after a certain number of steps (see Chap. 2).

Clearly, if 0 denotes the zero vector in $\{0,1\}^n$ then we have

$$\forall x \in \{0,1\}^n \qquad d(x,0)=x.$$

Moreover

$$\forall x \in \{0,1\}^n \qquad d(x,\tilde{x}^j)=e_j \qquad (j=1,2,\ldots,n).$$

Therefore, let F be a map of $\{0,1\}^n$ into itself. As always, our aim is to study the behaviour (in $\{0,1\}^n$) of the discrete iteration

$$x^{r+1}=F(x^r) \qquad (r=0,1,2,\ldots)$$

where x^0 is given in $\{0,1\}^n$.

Let us recall the formalism used. The relation $y=F(x)$ is developed as

$$y_i=f_i(x_1,\ldots,x_n) \qquad i=1,2,\ldots,n.$$

We also have (see Chap. 2) that

$$\forall x,y \in \{0,1\}^n \qquad d(F(x),F(y)) \le B(F)\,d(x,y)$$

since $B(F)$ denotes the incidence matrix for F.

We have already noted that such an inequality is rather global. In the following we therefore work more locally in order to obtain *a local analysis of the iteration in the neighbourhood of a fixed point or a cycle.*

Taking into account the type of convergence considered in $\{0,1\}^n$ (stationary convergence) we may define the notion of a *discrete derivative* of F at a vertex x of $\{0,1\}^n$.

Prior to this definition we note, however, that *all the operations in this chapter* (sum in $\{0,1\}^n$, matrix product ...) *are boolean operations* $(1+1=1)$. *Moreover, the inequality* $\le$ *used on* $\{0,1\}^n$ *is the componentwise inequality* (*for the ordering* $0\le 0\le 1\le 1$ *on* $\{0,1\}$).

Definition. The *discrete derivative* of F at a point x in $\{0,1\}^n$ is the boolean $n\times n$ matrix defined by

$$f_{ij}(x)=1 \quad \text{if } f_i(x_1,\ldots,x_j,\ldots,x_n)\neq f_i(x_1,\ldots,\bar{x}_j,\ldots,x_n)$$
$$f_{ij}(x)=0 \quad \text{otherwise.}$$

This discrete derivative is written as

$$F'(x)=(f_{ij}(x)).$$

(Stated differently: $f_{ij}(x)=1$ if $f_i(x)\neq f_i(\tilde{x}^j)$
$f_{ij}(x)=0$ otherwise
$i,j=1,2,\ldots,n$.)

It is clear that the data items x, $F(x)$ and $F'(x)$ define $F(y)$ for all y in the neighbourhood of x.

For studies about boolean derivatives, see Refs. [215]→[218].

Example

$$n=3 \quad x=(0\ 1\ 1) \quad F(x)=(1\ 0\ 1) \quad \text{and} \quad F'(x)=\begin{bmatrix}1 & 0 & 1\\ 1 & 0 & 0\\ 1 & 0 & 1\end{bmatrix}.$$

Then

$$\begin{array}{lcl} \tilde{x}^1=(1\ 1\ 1) & & F(\tilde{x}^1)=(0\ 1\ 0)\\ \tilde{x}^2=(0\ 0\ 1) & \text{and necessarily} & F(\tilde{x}^2)=(1\ 0\ 1)\\ \tilde{x}^3=(0\ 1\ 0) & & F(\tilde{x}^3)=(0\ 0\ 0). \end{array}$$

In particular, to say that $F'(x)=0$ (boolean zero matrix), is the same as to say that $F(y)$ is constant (and equal to $F(x)$) at all points y in the immediate neighbourhood of x.

We now establish a theorem that, although elementary, is interesting since it links the incidence matrix with the discrete derivative.

Theorem 1. *If $B(F)$ is the incidence matrix for F then*

a) $\forall x \in \{0,1\}^n \quad F'(x) \leq B(F)$.

b) $\operatorname{Sup}_{x \in \{0,1\}^n} \{F'(x)\} = B(F)$.

To prove a) it suffices to show that if $B(F)$ has a zero in position i, j ($b_{ij}=0$) then the same is true for $F'(x)$ independent of x.

Clearly, whenever f_i is by hypothesis independent of the j-th variable x_j, then we have necessarily for all $x \in \{0,1\}^n$ that

$$f_i(x)=f_i(\tilde{x}^j).$$

Since this means that $f_{ij}(x)=0$, we have proven a).

From a) it furthermore follows that

$$\sup_{x \in \{0,1\}^n} \{F'(x)\} \leq B(F).$$

(The sup and the $\leq$ are taken elementwise on the boolean matrices of size $n \times n$ using the order $0 \leq 0 \leq 1 \leq 1$ on $\{0,1\}$.)

In order to show that the above inequality is indeed an equality, and thus show b), it is sufficient to show that if $B(F)$ has a 1 in position i, j ($b_{ij}=1$) then there exists an x in $\{0,1\}^n$ such that $F'(x)$ also has a 1 in position i, j ($f_{ij}(x)=1$).

Now, to say that $b_{ij}=1$ is the same as to say that f_i actually depends on x_j. Therefore there exists an x in $\{0,1\}^n$ such that f_i evaluated at $x=(x_1, \ldots, X_j, \ldots, x_n)$ and $\tilde{x}^j=(x_1, \ldots, \bar{x}_j, \ldots, x_n)$ results in two *different* values, that is

$$f_i(x) + f_i(\tilde{x}^j)$$

from which $f_{ij}(x)=1$. □

2. The Discrete Derivative and the Vector Distance

Theorem 2.

$$\forall x \in \{0,1\}^n, \quad \forall y \in V_x \quad d(F(x), F(y)) = F'(x)\, d(x,y)$$

(boolean product of $F'(x)$ *by* $d(x,y)$*).*

The proof is elementary. If $y=x$ then the above relation is trivially verified. If not, then since y is an element from V_x it must be a certain $\tilde{x}^j$. Then $d(x,y)=e_j$ such that $F'(x)\,d(x,y)$ represents the j-th column of $F'(x)$ which by definition coincides with the vector $d(F(x), F(y))$. □

Thus the result of the above theorem is that if u and v are neighbours in $\{0,1\}^n$ (that is to say, they differ in at most one component) then

$$d(F(u), F(v)) = F'(u)\, d(u,v) = F'(v)\, d(u,v),$$

since $F'(u)$ and $F'(v)$ have the same j-th column whenever u and v only differs in the j-th component ($d(u,v)=e_j$).

We will now use the preceding theorem for a majorization of $d(F(x), F(y))$ valid for *any* x and y in $\{0,1\}^n$.

In this case, one may pass from x to y by a *chain* $[x, u_1, u_2, \ldots, u_{r-1}, y]$ denoted by $[x, y]$ where each point is a neighbour of the next point in the chain (belonging to its immediate neighbourhood).

Clearly, there are many different chains leading from x to y.

We will take chains with *minimal length* (which implies that $r \leq n$). Then

$$d(F(x), F(y)) \leq d(F(x), F(u_1)) + d(F(u_1), F(u_2)) + \ldots + d(F(u_{r-1}), F(y)).$$

↑ componentwise inequality in $\{0,1\}^n$ $(0 \leq 0 \leq 1 \leq 1)$

↑ ↑ ↑ boolean sum $(1+1=1)$ in $\{0,1\}^n$

$$= F'(x)\, d(x, u_1) + F'(u_1)\, d(u_1, u_2) + \ldots + F'(u_{r-1})\, d(u_{r-1}, y).$$

Since the chain $[x, y]$ has *minimal* length it follows that $d(x, u_1), d(u_1, u_2), \ldots, d(u_{r-1}, y)$ is formed by basis vectors in $\{0,1\}^n$, *all different*. We therefore obtain a majorization of $d(F(x), F(y))$ by a boolean sum of a certain column of $F'(x)$ with *one other* of $F'(u_1)$ and so on.

One may majorize in a more global manner by defining

$$M = \operatorname*{Sup}_{z \in [x,y]} \{F'(z)\}.$$

(This Sup is taken elementwise, moreover M depends on the particular choice of the chain $[x, y]$.) Then

$$d(F(x), F(y)) \leq M[d(x, u_1) + d(u_1, u_2) + \ldots + d(u_{r-1}, y)].$$

Each of the vectors in the square brackets is a basis vector of $\{0,1\}^n$. *They are all different*, from which we conclude that the content of the square brackets is *equal* to $d(x, y)$. From this we obtain

Theorem 3. *For all x, y in $\{0,1\}^n$ one has*

$$d(F(x), F(y)) \leq \operatorname*{Sup}_{z \in [x, y]} \{F'(z)\}\, d(x, y)$$

for an arbitrary chain $[x, y]$ of minimal length.

Remarks. (1) It also suffices to take

$$\operatorname{Sup}\{F'(x), F'(u_1), \ldots, F'(u_{r-1})\}$$

for M. This is also written as

$$\operatorname*{Sup}_{z \in [x \to y]} \{F'(z)\}.$$

We also have

$$\operatorname*{Sup}_{z \in [y \to x]} \{F'(z)\} = \operatorname{Sup}\{F'(y), F'(u_{r-1}) \ldots F'(u_1)\}.$$

Using this we obtain the following majorizations:

$$d(F(x), F(y)) \leq \operatorname*{Sup}_{z \in [x \to y]} \{F'(z)\}\, d(x, y)$$

and

$$d(F(x), F(y)) \leq \operatorname*{Sup}_{z \in [y \to x]} \{F'(z)\}\, d(x, y).$$

(2) For x and y, arbitrary in $\{0,1\}^n$ there does not in general exist a ξ in $\{0,1\}^n$ such that

$$d(F(x), F(y)) = F'(\xi)\, d(x, y).$$

In order to convince oneself of this, it suffices to examine the following example where $n=2$:

	x		$F(x)$	
a	0	0	1	0
b	0	1	1	1
c	1	0	1	1
d	1	1	0	1

Then for $x = b = (0\ 1)$ and $y = c = (1\ 0)$ we get

$$d(F(x), F(y)) = \begin{bmatrix} 0 \\ 0 \end{bmatrix}, \qquad d(x, y) = \begin{bmatrix} 1 \\ 1 \end{bmatrix}.$$

Now

$$F'(a) = \begin{bmatrix} 0 & 0 \\ 1 & 1 \end{bmatrix}, \quad F'(b) = \begin{bmatrix} 1 & 0 \\ 0 & 1 \end{bmatrix}, \quad F'(c) = \begin{bmatrix} 0 & 1 \\ 1 & 0 \end{bmatrix}, \quad F'(d) = \begin{bmatrix} 1 & 1 \\ 0 & 0 \end{bmatrix}.$$

In neither case do we have the zero matrix.

Theorem 3 plays the same useful role as the inequality

$$\|F(x)-F(y)\| \leq \sup_{z\in[x,y]} \|F'(z)\| \, \|x-y\|$$

used in normed vector spaces.

(3) Clearly one may once more majorize in the inequality used in Theorem 3

$$d(F(x),F(y)) \leq \underbrace{\sup_{z\in[x,y]} \{F'(z)\}}_{M} d(x,y)$$

the matrix M by the matrix $B(F)$ (Theorem 1) independent of x and y.

In this manner we rediscover the basic inequality

$$\forall x, y \in \{0,1\}^n \qquad d(F(x),F(y)) \leq B(F)\, d(x,y)$$

which we now understand better to be a "global" majorization as opposed to the above more *local* results.

Example. $n=3$ F is defined by the table

x			$F(x)$		
0	0	0	1	0	0
0	0	1	1	0	1
0	1	0	0	0	0
0	1	1	1	0	1
1	0	0	0	0	1
1	0	1	0	0	0
1	1	0	1	1	1
1	1	1	0	1	0

For $x=(0\ 1\ 1)$ we get $F'(x)=\begin{bmatrix}1 & 0 & 1\\ 1 & 0 & 0\\ 1 & 0 & 1\end{bmatrix}$

$$x=(0\ 1\ 1), \quad \tilde{x}^1=(1\ 1\ 1), \quad \tilde{x}^2=(0\ 0\ 1), \quad \tilde{x}^3=(0\ 1\ 0),$$

$$F(x)=(1\ 0\ 1), \quad F(\tilde{x}^1)=(0\ 1\ 0), \quad F(\tilde{x}^2)=(1\ 0\ 1), \quad F(\tilde{x}^3)=(0\ 0\ 0).$$

It is easily verified that

$$d(F(x),F(\tilde{x}^j)) = \underbrace{F'(x) \quad \underbrace{d(x\,\tilde{x}^j)}_{e_j}}_{\substack{j\text{-th column}\\ \text{of } F'(x)\\ (j=1,2,3).}}$$

Furthermore, for $y=(1\ 1\ 0)$ we get $F(y)=(1\ 1\ 1)$. A chain $[x, y]$ of minimal length is of the form $[x, u_1, y]$ where one may take $u_1=(1\ 1\ 1)$ [or $(0\ 1\ 0)$].

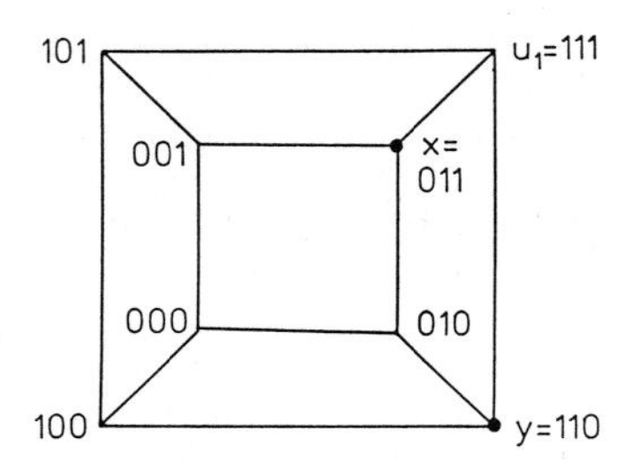

From this we get

$$d(F(x), F(y))=\begin{bmatrix}0\\1\\0\end{bmatrix} \leq \operatorname{Sup}\{F'(x), F'(u_1), F'(y)\}\, d(x, y).$$

Now

$$F'(x)=\begin{bmatrix}1&0&1\\1&0&0\\1&0&1\end{bmatrix}, \quad F'(u_1)=\begin{bmatrix}1&0&1\\1&1&0\\1&0&1\end{bmatrix}, \quad F'(y)=\begin{bmatrix}1&1&1\\1&1&0\\1&0&1\end{bmatrix},$$

from which we get

$$d(F(x), F(y))=\begin{bmatrix}0\\1\\0\end{bmatrix} \leq \begin{bmatrix}1&1&1\\1&1&0\\1&0&1\end{bmatrix}\begin{bmatrix}1\\0\\1\end{bmatrix}=\begin{bmatrix}1\\1\\1\end{bmatrix}.$$

In this example

$$\begin{aligned}
&f_1(x)=\bar{x}_1\,\bar{x}_2+\bar{x}_1\,x_2\,x_3+x_1\,x_2\,\bar{x}_3\\
&f_2(x)=x_1\,x_2\\
&f_3(x)=x_1\,\bar{x}_3+x_3\,\bar{x}_1 \qquad \text{(boolean notation)},
\end{aligned}$$

from which

$$B(F)=\begin{bmatrix}1&1&1\\1&1&0\\1&0&1\end{bmatrix}.$$

We may verify that for all z in $\{0, 1\}^3$

$$F'(z) \leq B(F)$$

and

$$\operatorname{Sup}\{F'(z)\}=B(F).$$

Moreover

$$\begin{aligned}
\operatorname{Sup}\{F'(x), F'(u_1), F'(y)\}&=B(F)\\
=\operatorname{Sup}\{F'(u_1), F'(y)\}&=\sup_{z\in[y\to x]}\{F'(z)\}.
\end{aligned}$$

However,

$$\operatorname*{Sup}_{z\in[x\to y]}\{F'(z)\}=\operatorname{Sup}\{F'(x),F'(u_1)\}=\begin{bmatrix}1&0&1\\1&1&0\\1&0&1\end{bmatrix}\leq B(F).$$

The inequality

$$d(F(x),F(y))\leq \operatorname{Sup}\{F'(x),F'(u_1)\}\;d(x,y)$$

however, may be written

$$\begin{bmatrix}0\\1\\0\end{bmatrix}\leq\begin{bmatrix}1&0&1\\1&1&0\\1&0&1\end{bmatrix}\begin{bmatrix}1\\0\\1\end{bmatrix}=\begin{bmatrix}1\\1\\1\end{bmatrix}$$

and it does not provide any more information than the trivial majorization obtained earlier.

3. Application: Characterization of the Local Convergence in the Immediate Neighbourhood of a Fixed Point

Definition. Let $\xi=F(\xi)$ be a fixed point of F in $\{0,1\}^n$. ξ is called *attractive in its immediate neighbourhood* if the following two conditions are valid:

a) $F(V_\xi)\subset V_\xi$.

b) For all x^0 from V_ξ, the iteration $x^{r+1}=F(x^r)$ (which remains in V_ξ according to a)) reaches ξ in at most n steps ($x^n=F^n(x^0)=\xi$).

Based on this definition we prove:

Theorem 4. *In order that ξ shall be attractive in its immediate neighbourhood it is necessary and sufficient that*

1) $F'(\xi)$ has at most one 1 in each column.

2) The boolean spectral radius of $F'(\xi)$ is zero (that is, there exists a permutation matrix P such that

$$P^t\,F'(\xi)\,P$$

is strictly lower triangular).

a) *The sufficient condition.* If 1) and 2) are satisfied then we have for all $x\in V_\xi$ that

$$d(F(x),\xi)=F'(\xi)\underbrace{d(x,\xi)}_{e_i}.$$

By assumption, the columns of $F'(\xi)$ may only be zero or a basis vector. This is the same as to say that all columns of $F'(\xi)$ belong to V_0, the immediate neighbourhood of the zero vector.

This means that $d(F(x), \xi) \in V_0$ from which it follows that $F(x) \in V_\xi$.

Then, for all x^0 in V_ξ, the sequence

$$x^{r+1} = F(x^r)$$

remains in V_ξ according to the preceding and we have that

$$d(x^r, \xi) = d(F(x^{r-1}), \xi) = F'(\xi)\, d(x^{r-1}, \xi) = \ldots = [F'(\xi)]^r\, d(x^0, \xi).$$

Since $\rho(F'(\xi)) = 0$ there exists an integer $p \leq n$ (Chap. 3, Theorem 5, p. 48) such that

$$[F'(\xi)]^p = 0 \qquad \text{(boolean power)},$$

and hence we have

$$d(x^p, \xi) = 0 \qquad \text{from which } x^p = \xi.$$

In other words the sequence reaches the fixed point ξ in at most p steps.

b) *The necessary condition.*

Suppose that ξ is attractive in V_ξ and that one of the columns of $F'(\xi)$, say the j-th, contains several 1's. We then have that

$$d(F(\tilde{\xi}^j), \xi) = \underbrace{F'(\xi)\, \underbrace{d(\tilde{\xi}^j, \xi)}_{e_j}}_{\substack{\text{the } j\text{-th column} \\ \text{of } F'(\xi).}}$$

This shows that $F(\tilde{\xi}^j)$ is *not* a neighbour of ξ since it differs from ξ in more than one component. Therefore F does not map V_ξ into itself, which is a contradiction with a).

We now show that the spectral radius of $F'(\xi)$ is necessarily zero.

From the assumptions, it follows that

$$\forall x \in V_\xi \qquad F^n(x) = \xi.$$

In particular for $x = \tilde{\xi}^j$ we get

$$d(F(\tilde{\xi}^j), \xi) = F'(\xi)\, e_j.$$

From the assumptions $F(\tilde{\xi}^j)$ is also in V_ξ, from which

$$\begin{aligned} d(F^2(\tilde{\xi}^j), \xi) &= F'(\xi)\, d(F(\tilde{\xi}^j), \xi) \\ &= [F'(\xi)]^2 \underbrace{d(\tilde{\xi}^j, \xi)}_{e_j} \end{aligned}$$

and by recurrence

$$d(F^n(\tilde{\xi}^j), \xi) = [F'(\xi)]^n e_j \qquad (j = 1, 2, \ldots, n).$$

From the assumptions, the first member is zero. Thus all columns of $[F'(\xi)]^n$ are zero. Then $[F'(\xi)]^n$ (boolean power) is the zero matrix from which

$$\rho(F'(\xi)) = 0.$$

Furthermore, $F'(\xi)$ is a strictly lower triangular matrix (up to a permutation of the rows and the same permutation of the columns). □

Example. $n=3 \quad \xi = (1\ 1\ 0)$ and $F'(\xi) = \begin{bmatrix} 0 & 0 & 0 \\ 0 & 0 & 1 \\ 1 & 0 & 0 \end{bmatrix}$.

$F'(\xi)$ satisfies conditions 1) and 2) of the preceding theorem. We then necessarily have that

	x	$F(x)$
	0 0 0	
	0 0 1	
$\tilde{\xi}^1$	0 1 0	1 1 1
	0 1 1	
$\tilde{\xi}^2$	1 0 0	1 1 0
	1 0 1	
ξ	1 1 0	1 1 0
$\tilde{\xi}^3$	1 1 1	1 0 0

This shows that a part of the iteration graph for F must be

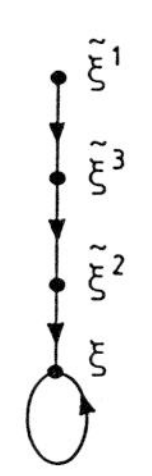

This clearly verifies that ξ is attractive in its immediate neighbourhood V_ξ.

Remark. The condition $\rho(F'(\xi)) = 0$ is by itself not sufficient to guarantee that ξ is attractive *in all of* V_ξ. It simply guarantees that one has one of the following schemes:

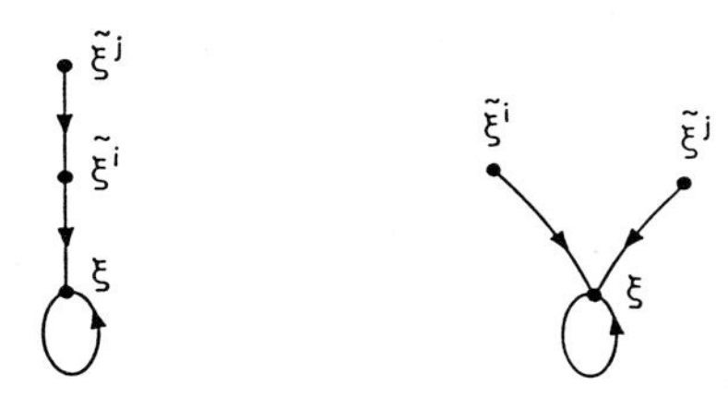

Indeed, there exists a permutation matrix P such that

$$P^t F'(\xi) P \quad \text{is strictly lower triangular.}$$

From this

$$P^t F'(\xi) P e_n = 0 \quad \text{and} \quad P^t F'(\xi) P e_{n-1} \leq e_n$$

and furthermore

$$F'(\xi) P e_n = 0 \quad \text{and} \quad F'(\xi) P e_{n-1} \leq P e_n.$$

We write e_i for $P e_n$ and e_j for $P e_{n-1}$ ($P e_n = e_i$ and $P e_{n-1} = e_j$) and we have first that

$$F'(\xi) e_i = 0.$$

This means that $d(F(\tilde{\xi}^i), \xi) = 0$ from which we have $F(\tilde{\xi}^i) = \xi$.

We now have

$$F'(\xi) e_j \leq e_i.$$

Only two cases are now possible:

- Either $F'(\xi) e_j = e_i$ from which $d(F(\tilde{\xi}^j), \xi) = e_i$ that is to say

$$F(\tilde{\xi}^j) = \tilde{\xi}^i.$$

This clearly implies that we have the following schema:

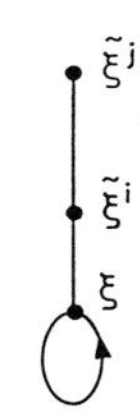

- Or

$$F'(\xi) e_j = 0$$

from which

$$d(F(\tilde{\xi}^j), \xi) = 0$$

or furthermore

$$F(\tilde{\xi}^j) = \xi.$$

This results in the following scheme:

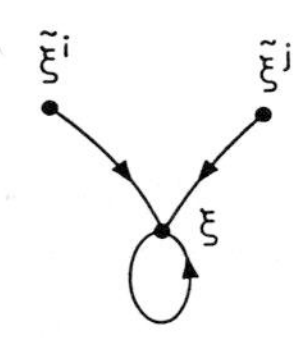

Example. $n=3$ $\xi=(1\ 1\ 0)$ and $F'(\xi)=\begin{bmatrix}0 & 1 & 1\\ 0 & 0 & 0\\ 0 & 1 & 0\end{bmatrix}$.

Clearly $\rho(F'(\xi))=0$ but $F'(\xi)$ has 2 ones in the second column. One therefore necessarily gets

	x	$F(x)$	
	0 0 0		
	0 0 1		
$\tilde{\xi}^1$	0 1 0	1 1 0	ξ
	0 1 1		
$\tilde{\xi}^2$	1 0 0	0 1 1	?
	1 0 1		
ξ	1 1 0	1 1 0	ξ
$\tilde{\xi}^3$	1 1 1	0 1 0	$\tilde{\xi}^1$

From which the following scheme is obtained:

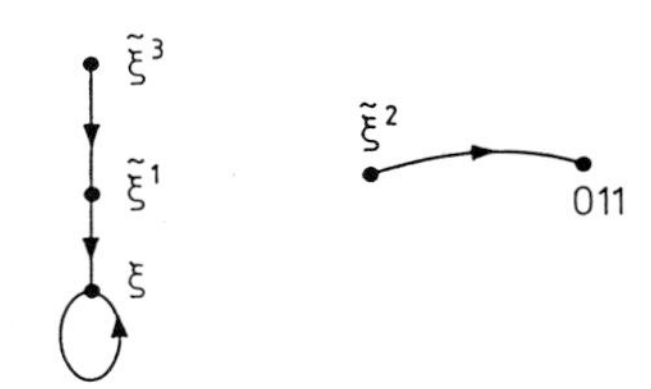

ξ is therefore not attractive in *all* of V_ξ since $F(\tilde{\xi}^2)\notin V_\xi$.

A very particular case. If $\xi=F(\xi)$ is a fixed point of F and $F'(\xi)=0$ (the boolean zero matrix) then the iteration graph for F in the immediate neighbourhood of ξ is the following:

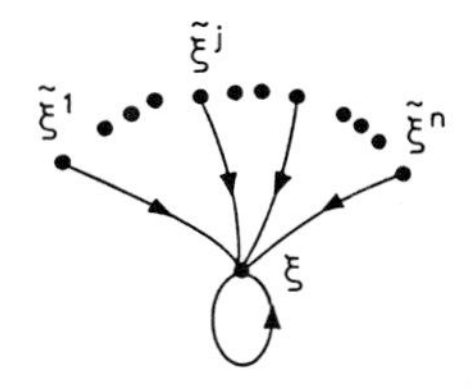

Otherwise stated

$$F(\tilde{\xi}^j)=\xi \qquad (j=1,2,\ldots,n)$$

because

$$d(F(\tilde{\xi}^j),\xi)=F'(\xi)\,e_j=0.$$

4. Interpretation in Terms of Automata Networks

The notions of an immediate neighbourhood and of an attractive fixed point in an immediate neighbourhood may be given good interpretations in the context of automata networks.

Therefore let $x=(x_1, \ldots, x_n)$ be a configuration where each cell i of the network is in the state $x_i \in \{0, 1\}$ $(i=1, 2, \ldots, n)$.

A *neighbouring configuration of* x is then a configuration $\tilde{x}^j$ obtained from x by only modifying the j-th component of x $(j=1, 2, \ldots, n)$.

A fixed point of F *is a stable configuration* ξ. *To say that* ξ *is attractive in its immediate neighbourhood is to say that one single cell of* ξ *is to be modified after which the automata network is made to work in parallel on this modified configuration. Then* ξ *is regained while only passing through neighbouring configurations of* ξ (in at most n steps).

We illustrate this using the example from p. 105. The stable configuration $\xi=(1\ 1\ 0)$ may be visualized in the following manner:

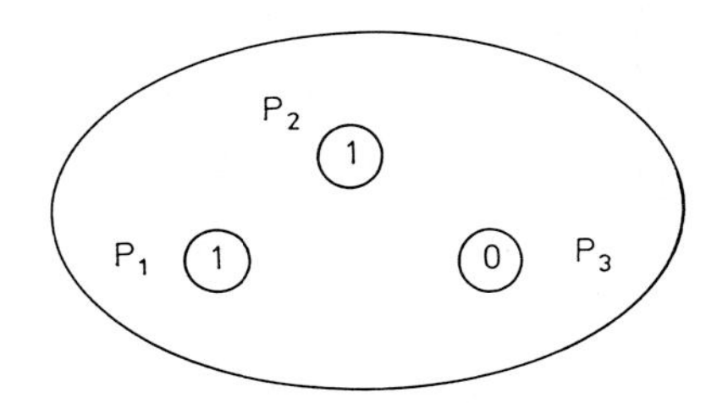

Without specifying the connectivity graph between the cells (it is not completely determined) one may visualize the parallel iteration in the immediate neighbourhood of ξ as follows:

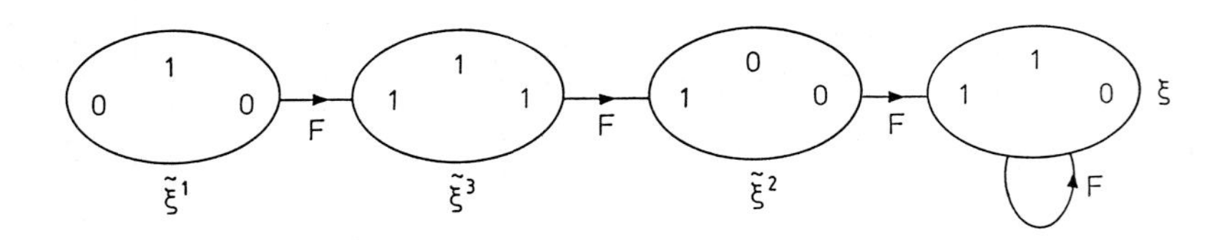

5. Application: Local Convergence in a Massive Neighbourhood of a Fixed Point

We will now study the convergence of the sequence $x^{r+1}=F(x^r)$ *towards a fixed point starting from a neighbourhood possibly larger than the immediate neighbourhood.* For this we need the notion of a *massive neighbourhood* in $\{0, 1\}^n$.

Definition. Let x be a point in $\{0, 1\}^n$ and let V be a subset of $\{0, 1\}^n$ containing x with the property that if $u \in V$ then all $v \in \{0, 1\}^n$ satisfying $d(x, v) \leq d(x, u)$ are also in V. Then V is called a *massive neighbourhood* of x.

Stated differently: If $u \in V$ then all v in $\{0, 1\}^n$ that are "closer to x than u" in the sense of the vector distance d are also in V.

To a massive neighbourhood V of x there is also an associated massive neighbourhood of 0 (vector 0) which is the set

$$W = \{d(x, u) \ \forall u \in V\}.$$

Clearly 0 is in W. It now follows that W is a massive neighbourhood of 0 since if α is in W then for all $\beta \in \{0, 1\}^n$ satisfying

$$\beta \leq \alpha \qquad \text{(also written } d(0, \beta) \leq d(0, \alpha))$$

we have

1) the existence of a u in V such that $d(x, u) = \alpha$

2) the existence of a v in $\{0, 1\}^n$ such that $d(x, v) = \beta$,

clearly satisfying $d(x, v) \leq d(x, u)$ from the above discussion. This means that $v \in V$ which clearly implies that $\beta \in W$.

Example. $n = 3$ with

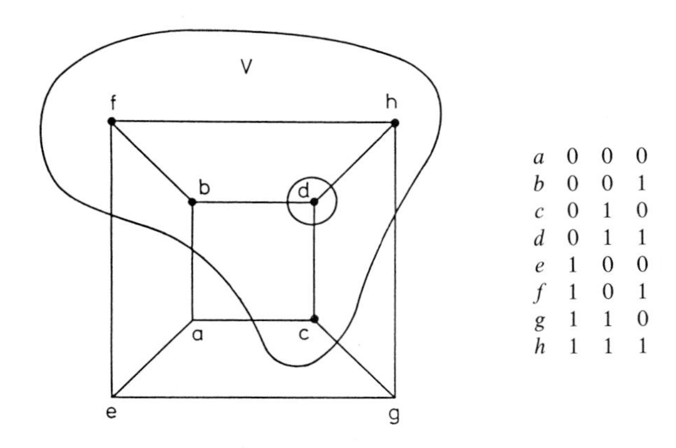

a	0	0	0
b	0	0	1
c	0	1	0
d	0	1	1
e	1	0	0
f	1	0	1
g	1	1	0
h	1	1	1

$V = \{d, b, c, f, h\}$ is a massive neighbourhood of d. The associated massive neighbourhood W of $a = (0\ 0\ 0)$ is

$$W = \{a, c, b, g, e\}.$$

Definition. One says that a fixed point $\xi = F(\xi)$ of F is *attractive in a massive neighbourhood* V of ξ if again

a) $F(V) \subset V$.

b) For all x^0 from V, the iteration $x^{r+1} = F(x^r)$ (which remains in V from a)) becomes stationary at ξ in at most n steps.

Below we are stating *sufficient* conditions for a fixed point ξ to be attractive in a given massive neighbourhood.

Theorem 5. *Let V be a massive neighbourhood of a fixed point ξ of F and let $W=\{d(\xi, x)\ x\in V\}$ be the associated massive neighbourhood of 0. Furthermore, let*

$$M=\operatorname*{Sup}_{z\in V}\{F'(z)\}.$$

If the following conditions are satisfied:

$$C\begin{cases} MW\subset W \\ \rho(M)=0 \qquad \textit{(boolean spectral radius)}\end{cases}$$

then ξ is attractive in V.

1) We first show that under the conditions C, F maps V into itself.
Therefore let x be in V. Then

$$d(\underbrace{F(\xi)}_{\xi}, F(x))\leq d(F(\xi), F(u_1))+\dots d(F(u_{r-1}), F(x))$$

where the chain $[\xi, x]=\{\xi, u_1, \dots, u_{r-1}, x\}$, of minimal length, is formed by taking points that pairwise are neighbours in V (because V is massive it follows that all the u_i are in V since

$$d(\xi, u_i)\leq d(\xi, x) \quad \text{with } x\in V).$$

From this it follows that:

$$\begin{aligned} d(\xi, F(x)) &\leq F'(\xi)\, d(\xi, u_1)+\dots F'(u_{r-1})\, d(u_{r-1}, x) \\ &\leq M[d(\xi, u_1)+\dots d(u_{r-1}, x)]. \end{aligned}$$

Once more, therefore, we have a chain $[\xi, x]$ of minimal length, where all the terms in the square brackets are basis vectors from $\{0, 1\}^n$, *all different*. This means that the result in the square brackets is *equal* to $d(\xi, x)$.
Finally, we get

$$\forall x\in V \qquad d(F(x), \xi)\leq M\, d(x, \xi).$$

Now x is in V which means that $d(x, \xi)$ belongs to W. Since $MW\subset W$, it furthermore follows that $v=M\,d(x, \xi)$ is in W.
We therefore have that

$$d(F(x), \xi)\leq v \qquad \text{with } v\in W.$$

Now, since W is massive, it follows that $d(F(x), \xi)$ belongs to W, that is to say, finally, that $F(x)$ is in V, from which the result follows.

2) We now show that

$$\forall x\in V \qquad F^n(x)=\xi.$$

Indeed, for all $x=x^0$ in V the successive iterates $x^{r+1}=F(x^r)$ remain in V after the preceding discussion and we have

$$\begin{aligned} d(x^1,\xi) &\leq M\, d(x^0,\xi)\\ d(x^2,\xi) &\leq M^2\, d(x^0,\xi)\\ &\vdots\\ d(x^n,\xi) &\leq M^n\, d(x^0,\xi). \end{aligned}$$

Now, to say that $\rho(M)=0$ is the same as to say $M^n=0$, from which $x^n=\xi$. □

Example. $n=3$ $\xi=a=(0\ 0\ 0)$ $V=W=\{a,b,c,e,f\}$ is a massive neighbourhood of a.

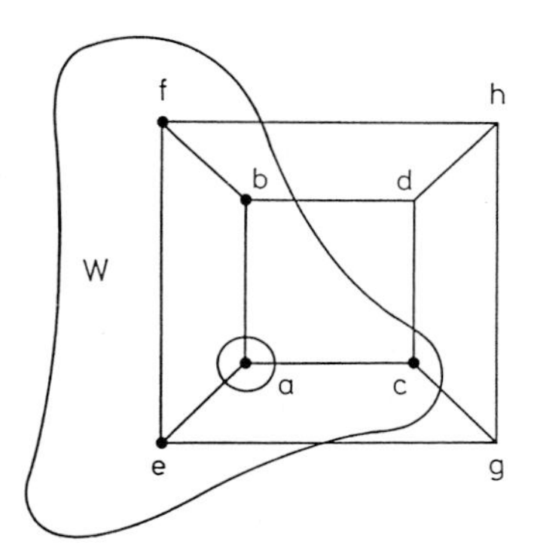

F is constructed starting with a (fixed point) and

$$F'(a)=\begin{bmatrix}0&1&0\\0&0&0\\1&1&0\end{bmatrix},\quad F'(b)=\begin{bmatrix}0&1&0\\0&0&0\\1&1&0\end{bmatrix},\quad F'(c)=\begin{bmatrix}0&1&0\\0&0&0\\0&1&0\end{bmatrix},$$

$$F'(e)=\begin{bmatrix}0&1&0\\0&0&0\\1&0&0\end{bmatrix},\quad F'(f)=\begin{bmatrix}0&0&0\\0&0&0\\1&0&0\end{bmatrix},$$

from which

	x	$F(x)$	
a	0 0 0	0 0 0	a
b	0 0 1	0 0 0	a
c	0 1 0	1 0 1	f
d	0 1 1	1 0 1	f
e	1 0 0	0 0 1	b
f	1 0 1	0 0 1	b
g	1 1 0	1 0 1	f
h	1 1 1	0 0 1	b

Here $M=\text{Sup}\{F'(a),F'(b),F'(c),$

$$F'(e),F'(f)\}=\begin{bmatrix}0&1&0\\0&0&0\\1&1&0\end{bmatrix}.$$

We now have $MW = \{a, b, f\} \subset W = \{a, b, c, e, f\}$ and $\rho(M) = 0$ (up to a permutation, M is a strictly lower triangular matrix).

We may easily verify the local convergence towards a in the massive neighbourhood $M = \{a, b, c, e, f\}$ since the iteration graph of F is the following:

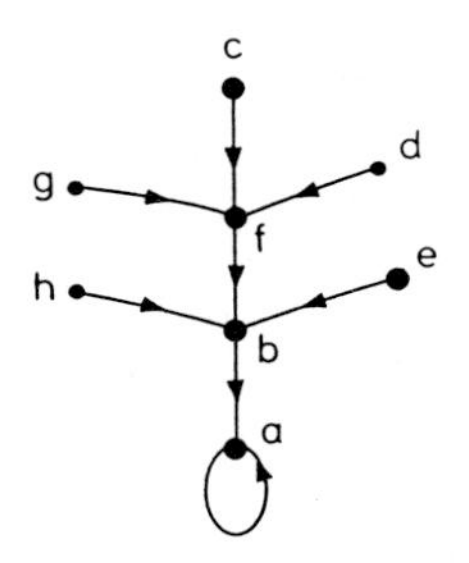

(This is the example considered on p. 65, illustrating the fact that even though F is not contracting its iteration graph is still simple.)

Remarks. (1) *The notion of local convergence in a massive neighbourhood of a fixed point* ξ *may also be interpreted in the context of automata networks* where the fixed point ξ is a stable configuration. *If one starts from a configuration* x^0 *belonging to a massive neighbourhood* V *of* ξ, *and if the automata are made to operate in parallel, then it will return to* ξ *in no more than* n *steps, each step producing a configuration belonging to* V.

Let us illustrate this using the preceding example. We get

$$f_1(x) = \bar{x}_1 x_2 + x_1 x_2 \bar{x}_3$$
$$f_2(x) = 0$$
$$f_3(x) = x_1 + x_2$$

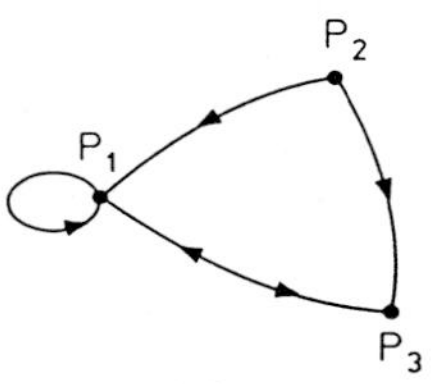

Connectivity graph for the network (3 cells)

Furthermore, we have the stable configuration

0

0 0

the massive neighbourhood $V=\{0\,{}^{0}\,0,\ 0\,{}^{0}\,1,\ 0\,{}^{1}\,0,\ 1\,{}^{0}\,0,\ 1\,{}^{0}\,1\}$ and a part of the iteration graph for F

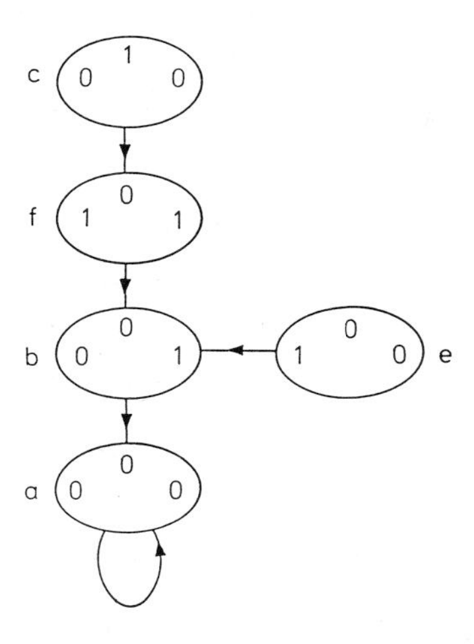

(2) Theorem 5 only gives *sufficient* conditions for convergence in V (this is in contrast to Theorem 4 which gave both *necessary and sufficient* conditions). The reason for this is the difference between the equality and the majorization used. Moreover, when recalling the proof of Theorem 5 we realize that we have majorized $F'(z)$ a bit too much by taking the Sup over all the elements of V (see also Remark 1 to Theorem 3). It is really sufficient to take the Sup on a subset of V that must

"be a union of chains of minimal length starting from the fixed point ξ, visiting all the points of V, and not including the terminal elements."

Let us denote such a set by $\hat{V}$ (it is not defined uniquely since in the preceding example one may take $\hat{V}=\{a,b\}$ or $\hat{V}=\{a,e\}$).

$\hat{V}$ is a subset of V since V is massive.

It is now sufficient to take

$$\hat{M}=\operatorname*{Sup}_{z\in\hat{V}}\{F'(z)\}\leq M$$

in order to obtain sufficient conditions for guaranteeing local convergence in V:

$$\mathrm{C}^*\ \begin{cases}\hat{M}W\subset W\\ \rho(\hat{M})=0.\end{cases}$$

The conditions C* are somewhat less restrictive than the conditions C of Theorem 5. Indeed, since $\hat{M}\leq M$, we have under conditions C that

$$\hat{M}W\leq MW \quad \text{(obvious notation).}$$

Now $MW\subset W$ and W is massive, therefore $\hat{M}W\subset W$.

Moreover, from the boolean Perron-Frobenius theorem we get that $\hat{M} \leq M$ implies $\rho(\hat{M}) \leq \rho(M)$. Now from the assumptions $\rho(M)=0$ we get $\rho(\hat{M})=0$. We therefore have established that

$$C \Rightarrow C^*.$$

In the preceding example we have, moreover, that for

$$\hat{V}=\{a,b\} \qquad \hat{M}=\operatorname{Sup}\{F'(a),F'(b)\}=\begin{bmatrix}0&1&0\\0&0&0\\1&1&0\end{bmatrix}=M$$

and for

$$\hat{V}=\{a,e\} \qquad \hat{M}=\operatorname{Sup}\{F'(a),F'(e)\}=\begin{bmatrix}0&1&0\\0&0&0\\1&1&0\end{bmatrix}=M.$$

(3) Counterexample for Theorem 5: $n=3 \quad \xi=c=(0\ 1\ 0)$

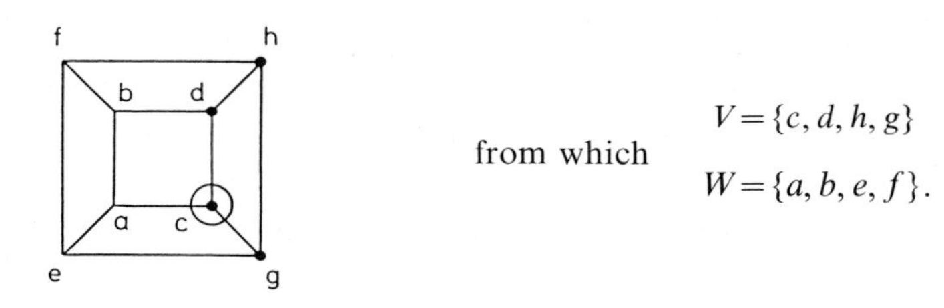

We impose a strictly lower triangular form on $F'(c)$ and $F'(d)$

$$F'(c)=\begin{bmatrix}0&0&0\\1&0&0\\1&1&0\end{bmatrix}, \qquad F'(d)=\begin{bmatrix}0&0&0\\1&0&0\\1&0&0\end{bmatrix}$$

	x	$F(x)$	
a	0 0 0	0 1 1	d
b	0 0 1	0 1 0	c
c	0 1 0	0 1 0	c
d	0 1 1	0 1 0	c
e	1 0 0	0 0 1	b
f	1 0 1	1 0 1	f
g	1 1 0	0 0 1	b
h	1 1 1	0 0 1	b

We may take $\hat{V}=\{c,d\}$. (We might also have taken $\hat{V}=\{c,g\}$.) Here

$$\hat{M}=\operatorname{Sup}\{F'(c),F'(d)\}=\begin{bmatrix}0&0&0\\1&0&0\\1&1&0\end{bmatrix}$$

$\rho(\hat{M})=0$ but $\hat{M}W=\{a,d\} \not\subset W$.

The conditions C* are *not* verified. Similarly, and even more so, the conditions C are also not verified.

One finds here that one does not have local convergence towards c in V since $F(g)=b \notin V$.

In this example the iteration graph for F is the following:

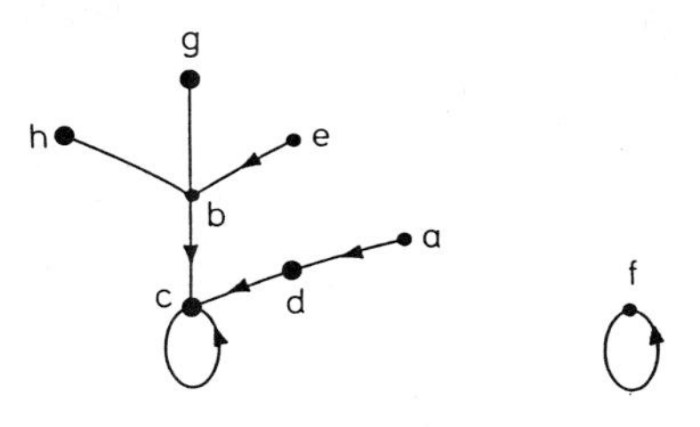

where $V=\{c, d, h, g\}$.

(4) *Case where* $V=V_\xi$

If $V=V_\xi$ (the immediate neighbourhood of a fixed point ξ which is also clearly a massive neighbourhood of ξ) then W (the immediate neighbourhood of 0) is formed by the zero vector and the set of basis vectors in $\{0, 1\}^n$ and $\hat{M}$ is reduced to $F'(\xi)$. The conditions C* are then those of Theorem 4 (which are, as we recall, *necessary* and sufficient).

(5) *Case where* $V=\{0, 1\}^n$

This is clearly a massive neighbourhood for all its elements. We therefore have $W=V=\{0, 1\}^n$ and

$$M= \operatorname*{Sup}_{z \in \{0, 1\}^n} \{F'(z)\} = B(F) \quad \text{(Theorem 1)}.$$

The condition $MW \subset W$ is clearly verified.

The preceding theorem may then be written as follows:

"If ξ is a fixed point for F and if $\rho(B(F))=0$ then for all x^0 in $\{0, 1\}^n$, the sequence $x^{r+1}=F(x^r)$ will reach ξ in no more than n steps and then remain stationary."

We therefore rediscover the contraction result (Chap. 4). Indeed, in this case, the existence (and unicity) of the fixed point is a *consequence* of the contraction condition $\rho(B(F))=0$.

6. Gauss-Seidel

We saw (Chap. 4) that whenever F is contracting on $\{0, 1\}^n$ (that is to say, whenever $\rho(B(F))=0$) then the associated Gauss-Seidel operator is also contracting. Furthermore, in this case,

$$B(G) \leq [I+L+\ldots+L^{n-1}]\, U$$

where L and U are defined by

$$B(F)=\begin{bmatrix} \ddots & U \\ L & \ddots \end{bmatrix}=L+U.$$

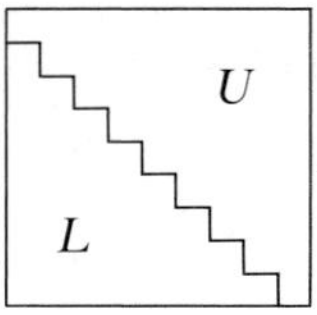

We therefore ask the following natural question: Suppose a fixed point ξ of F is attractive, for example in its immediate neighbourhood. Is that same fixed point attractive for G in its immediate neighbourhood? Moreover, does one have

$$G'(\xi)\leq[I+K+\ldots+K^{n-1}]\,T$$

with

$$F'(\xi)=\begin{bmatrix} \ddots & T \\ K & \ddots \end{bmatrix}\;?$$

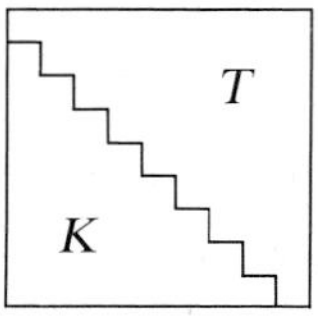

The question is answered negatively, as the following example shows:

	x	$F(x)$	$G(x)$
	0 0 0	0 1 0	0 1 0
	0 0 1	1	
ξ	0 1 0	0 1 0	0 1 0
	0 1 1	0 0 0	0 0 1
	1 0 0		
	1 0 1		
	1 1 0	0 0 0	0 1 0
	1 1 1		

(the elements not defined in the table for F are arbitrary)

$$F'(\xi)=\begin{bmatrix}0&0&0\\1&0&1\\0&0&0\end{bmatrix}=\underbrace{\begin{bmatrix}0&0&0\\1&0&0\\0&0&0\end{bmatrix}}_{K}+\underbrace{\begin{bmatrix}0&0&0\\0&0&1\\0&0&0\end{bmatrix}}_{T}.$$

We verify (Theorem 4) that ξ is attractive (for F) in its immediate neighbourhood.

We get

$$G'(\xi)=\begin{bmatrix}0&0&0\\0&0&1\\0&0&1\end{bmatrix}.$$

Now $\rho(G'(\xi))=1$ and ξ is not attractive for G in its immediate neighbourhood. Moreover $G(0\ 1\ 1)=(0\ 0\ 1)$, which is no longer in the immediate neighbourhood of ξ. Furthermore, we have

$$G'(\xi)\geq[I+K+K^2]\,T=\begin{bmatrix}0&0&0\\0&0&1\\0&0&0\end{bmatrix}.$$

7. The Derivative of a Function Composition

Let E and F be two operators on $\{0,1\}^n$ and form their composition

$$H=E\circ F.$$

Does one now have (as in the continuous case)

$$H'(x)=E'(F(x))\,F'(x)?$$

This is unfortunately not true, as the following simple example shows:

$n=2$

x	$F(x)$	$E(x)$	$H(x)$
0 0	0 1	0 1	1 1
0 1	1 1	1 1	0 0
1 0	1 0	1 1	1 1
1 1	1 0	0 0	1 1

Then, for example, for $x=(0\ 0)$, we obtain $F(x)=(0\ 1)$.

$$H'(x)=\begin{bmatrix}0&1\\0&1\end{bmatrix},\quad E'(F(x))=\begin{bmatrix}1&1\\1&0\end{bmatrix},\quad F'(x)=\begin{bmatrix}1&1\\1&0\end{bmatrix}$$

and

$$E'(F(x))\,F'(x)=\begin{bmatrix}1&1\\1&1\end{bmatrix}\neq H'(x).$$

In the following we establish various majorizations for $H'(x)$ in the general case:

For all $x\in\{0,1\}^n$ and all j between 1 and n, we have

$$d(H(x)\,H(\tilde{x}^j))=\underbrace{H'(x)\,\underbrace{d(x,\tilde{x}^j)}_{e_j}}_{\substack{j\text{-th column}\\ \text{of } H'(x)}}.$$

From the definition of H we get

$$d(H(x), H(\tilde{x}^j)) = d(E \circ \underbrace{F(x)}_{y}, E \circ \underbrace{F(\tilde{x}^j)}_{z})$$

$$\leq \operatorname*{Sup}_{t \in [y \to z]} \{E'(t)\}\, d(y, z)$$

(see Theorem 3, Remark 1). With the aid of the following relation

$$d(y, z) = d(F(x), F(\tilde{x}^j)) = F'(x)\, \underbrace{d(x, \tilde{x}^j)}_{e_j}$$

we get the first majorization of an arbitrary column of $H'(x)$ as:

(I) $$\underbrace{H'(x)\, e_j}_{\substack{j\text{-th column} \\ \text{of } H'(x)}} \leq \operatorname*{Sup}_{t \in [F(x) \to F(\tilde{x}^j)]} \{E'(t)\}\, \underbrace{F'(x)\, e_j}_{\substack{j\text{-th column} \\ \text{of } F'(x)}} \quad (j = 1, 2, \ldots, n)$$

This majorization clearly depends on the choice of the chain (of minimal length) $[F(x) \to F(\tilde{x}^j)]$.

Again, let V_x denote the immediate neighbourhood of x and let $F(V_x)$ be the image of V_x by F. $F(x)$ is clearly a member of $F(V_x)$, however, $F(V_x)$ is not necessarily a massive neighbourhood of $F(x)$ (see example below). Let us therefore define $\widehat{F(V_x)}$ as (see Theorem 5, Remark 2)

"A union of chains, of minimal length, starting from $F(x)$, visiting all the points of $F(V_x)$, and not including the terminal elements."

Here, $\widehat{F(V_x)}$ is not necessarily massive and $\widehat{F(V_x)}$ is therefore not necessarily a subset of $F(V_x)$ (see example below). Furthermore, $\widehat{F(V_x)}$ is not necessarily defined in a unique manner.

It is clear, though, that the majorization (I) results in

$$H'(x)\, e_j \leq \underbrace{\operatorname*{Sup}_{t \in \widehat{(V_x)}} \{E'(t)\}\, F'(x)\, e_j}_{\substack{\text{independent} \\ \text{of } j}}$$

This therefore implies that

(II) $$H'(x) \leq \operatorname*{Sup}_{t \in \widehat{F(V_x)}} \{E'(t)\}\, F'(x)$$

and finally the more coarse majorization

$$\text{(III)} \qquad H'(x) \leq \sup_{t\in\{0,1\}^n} \{E'(t)\}\, F'(x) = B(E)\, F'(x)$$

These three majorizations are now explained by an example.

$n=3 \qquad E=F \qquad H=F^2$ with

	x			$F(x)=E(x)$			$H(x)=F^2(x)$		
a	0	0	0	0	1	1	0	1	1
b	0	0	1	1	0	0	0	1	0
c	0	1	0	0	0	0	0	1	1
d	0	1	1	0	1	1	0	1	1
e	1	0	0	0	1	0	0	0	0
f	1	0	1	1	0	1	1	0	1
g	1	1	0	0	1	0	0	0	0
h	1	1	1	1	0	1	1	0	1

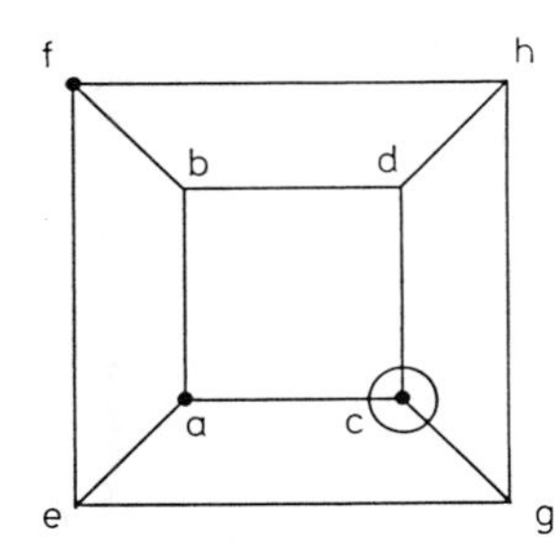

Let us take $x=(1\ 1\ 0)=g$ from which $F(x)=c$. With

$$V_x=\{g, c, e, h\}$$

and

$$F(V_x)=\{a, c, f\}$$

we get

$$H'(x)=\begin{bmatrix}0 & 0 & 1\\ 1 & 0 & 0\\ 1 & 0 & 1\end{bmatrix} \quad \text{and} \quad F'(x)=\begin{bmatrix}0 & 0 & 1\\ 1 & 0 & 1\\ 0 & 0 & 1\end{bmatrix}.$$

Majorization (I)

$j=1 \quad \tilde{x}^1=(0\ 1\ 0)=c \qquad F(\tilde{x}^1)=(0\ 0\ 0)=a$

the chain $[F(x)\to F(\tilde{x}^1)]$ reduces to $\{c\}$ and $F'(c)=\begin{bmatrix}0 & 0 & 0\\ 1 & 1 & 1\\ 0 & 1 & 1\end{bmatrix}$.

From this

$$\begin{bmatrix}0\\1\\1\end{bmatrix} = H'(x)\, e_1 \leq F'(c)\, \underbrace{F'(x)\, e_1} = \begin{bmatrix}0 & 0 & 0\\ 1 & 1 & 1\\ 0 & 1 & 1\end{bmatrix}\begin{bmatrix}0\\1\\0\end{bmatrix} = \begin{bmatrix}0\\1\\1\end{bmatrix}$$

and the inequality is here an equality.

$j=2 \quad \tilde{x}^2=(1\ 0\ 0)=e \qquad F(\tilde{x}^2)=(0\ 1\ 0)=c$

$[F(x)\to F(\tilde{x}^2)]$ reduces to $\{c\}$.

From this

$$\begin{bmatrix}0\\0\\0\end{bmatrix} = H'(x)\,e_2 \leq F'(c)\,F'(x)\,e_2 = \begin{bmatrix}0&0&0\\0&1&1\\0&1&1\end{bmatrix}\begin{bmatrix}0\\0\\0\end{bmatrix} = \begin{bmatrix}0\\0\\0\end{bmatrix}$$

and the inequality is an equality here.

$j=3 \quad \tilde{x}^3=(1\ 1\ 1)=h \qquad F(\tilde{x}^3)=(1\ 0\ 1)=f$

$[F(x)\rightarrow F(\tilde{x}^3)]$ may be taken equal to $\{c, a, b\}$ (one might also take $\{c, a, e\}$ or $\{c, d, b\}$).

With

$$\mathrm{Sup}(F'(c), F'(a), F'(b)) = \mathrm{Sup}\left\{\begin{bmatrix}0&0&0\\1&1&1\\0&1&1\end{bmatrix}, \begin{bmatrix}0&0&1\\0&1&1\\1&1&1\end{bmatrix}, \begin{bmatrix}0&1&1\\0&1&1\\1&1&1\end{bmatrix}\right\} = \begin{bmatrix}0&1&1\\1&1&1\\1&1&1\end{bmatrix}$$

one obtains

$$\begin{bmatrix}1\\0\\1\end{bmatrix} = H'(x)\,e_3 \leq \begin{bmatrix}0&1&1\\1&1&1\\1&1&1\end{bmatrix}\begin{bmatrix}1\\1\\1\end{bmatrix} = \begin{bmatrix}1\\1\\1\end{bmatrix}.$$

From this we finally have the majorization of $H'(x)$ as

$$H'(x) = \begin{bmatrix}0&0&1\\1&0&0\\1&0&1\end{bmatrix} \leq \begin{bmatrix}0&0&1\\1&0&1\\1&0&1\end{bmatrix}.$$

Majorization (II)

$x=g$

$V_x=\{g, c, e, h\}$ is the immediate neighbourhood of g

$F(V_x)=\{c, a, f\}$ is not a massive neighbourhood of $F(g)=c$.

We may then take $\widehat{F(V_x)}=\{c, a, b\}$. (We might equally take $\{c, a, e\}$.) Then

$$F'(c) = \begin{bmatrix}0&0&0\\1&1&1\\0&1&1\end{bmatrix}, \quad F'(a) = \begin{bmatrix}0&0&1\\0&1&1\\1&1&1\end{bmatrix}, \quad F'(b) = \begin{bmatrix}0&1&1\\0&1&1\\1&1&1\end{bmatrix}$$

$$\mathrm{Sup}\{F'(c), F'(a), F'(b)\} = \begin{bmatrix}0&1&1\\1&1&1\\1&1&1\end{bmatrix}$$

and the inequality II is written as

$$\begin{bmatrix}0&0&1\\1&0&0\\1&0&1\end{bmatrix}=H'(x)\leq\begin{bmatrix}0&1&1\\1&1&1\\1&1&1\end{bmatrix}\begin{bmatrix}0&0&1\\1&0&1\\0&0&1\end{bmatrix}=\begin{bmatrix}1&0&1\\1&0&1\\1&0&1\end{bmatrix}.$$

Majorization (III)

$$f_1(x)=x_3[x_1+\bar{x}_1\bar{x}_2]$$
$$f_2(x)=\bar{x}_1\bar{x}_2\bar{x}_3+\bar{x}_1x_2x_3+x_1\bar{x}_3$$
$$f_3(x)=\bar{x}_1\bar{x}_2\bar{x}_3+\bar{x}_1x_2x_3+x_1\bar{x}_2x_3+x_1x_2x_3$$

from which

$$B(F)=\begin{bmatrix}1&1&1\\1&1&1\\1&1&1\end{bmatrix}.$$

The inequality III

$$H'(x)\leq B(E)\,F'(x)$$

is written here as (where $E=F$)

$$\begin{bmatrix}0&0&1\\1&0&0\\1&0&1\end{bmatrix}=H'(x)\leq\begin{bmatrix}1&1&1\\1&1&1\\1&1&1\end{bmatrix}\begin{bmatrix}0&0&1\\1&0&1\\0&0&1\end{bmatrix}=\begin{bmatrix}1&0&1\\1&0&1\\1&0&1\end{bmatrix}.$$

8. The Study of Cycles: Attractive Cycles

Let F always be a map of $\{0,1\}^n$ into itself and let $a_0, a_1, \ldots, a_{r-1}$ be a *cycle of length* r for F (see Chap. 1). This is a set of r *distinct* points in $\{0,1\}^n$ such that

$$F(a_0)=a_1,\quad F(a_1)=a_2,\ldots,F(a_{r-2})=a_{r-1},\quad F(a_{r-1})=a_0.$$

Such a cycle is denoted $C=\{a_0, a_1, \ldots, a_{r-1}\}$. The immediate neighbourhood of a_i is still denoted V_{a_i} $(i=0,1,\ldots,r-1)$.

The immediate neighbourhood of C may be defined as the union of the V_{a_i} $(i=0,1,\ldots,r-1)$.

Definition. *The cycle* $C=\{a_0,\ldots,a_{r-1}\}$ *is said to be attractive in its immediate neighbourhood* if:

a) $F(V_{a_0})\subset V_{a_1}$; $F(V_{a_1})\subset V_{a_2}$ ... $F(V_{a_{r-2}})\subset V_{a_{r-1}}$ and $F(V_{a_{r-1}})\subset V_{a_0}$.

b) For all x^0 taken from one of the immediate neighbourhoods V_{a_i}, the iteration $x^{s+1}=F(x^s)$ (which runs through the V_{a_j} according to a)) becomes stationary at a_i in at most rp steps where $p \leq n$, that is

$$\exists p \quad (0 \leq p \leq n): \qquad x^{rp} = F^{rp}(x^0) = a_i.$$

Remarks. (1) It is clear that when the iteration has reached an element of the cycle then its runs through this cycle indefinitely. Moreover, the preceding definition is consistent (all points of the sequence may be considered to be x^0).

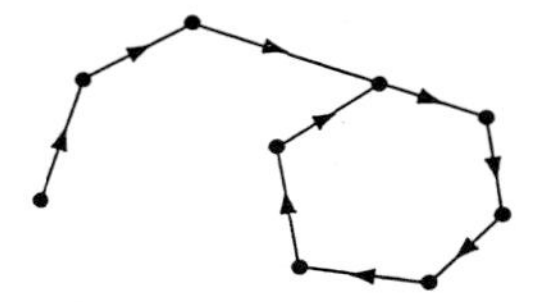

(2) According to the above definition, if the cycle C is attractive in its immediate neighbourhood then the V_{a_i} are necessarily mutually disjoint.*

Indeed, if $V_{a_i} \cap V_{a_j}$ were nonempty for a distinct pair i and j, then let x^0 be a member of this set. The iteration $x^{s+1} = F(x^s)$ then goes as follows:

$$F^{rp}(x^0) = x^{rp} = a_i$$
$$F^{rp'}(x^0) = x^{rp'} = a_j \qquad \text{with } p \text{ and } p' \leq n.$$

According to the preceding remark we then have

$$x^{rp+rp'} = a_i = a_j, \qquad \text{which is impossible.}$$

(3) *All elements a_i of the cycle C (of length r) for F are evidently fixed points for F^r. A consequence of the preceding definition is therefore that if C is attractive in its immediate neighbourhood (for F) then all elements a_i from C are attractive fixed points (for F^r) in their immediate neighbourhoods* (see example below).

We now establish a *characterization* for a cycle to be attractive in its immediate neighbourhood.

Theorem 6. *In order that a cycle $C = \{a_0, a_1, \ldots, a_{r-1}\}$ of F shall be attractive in its immediate neighbourhood it is necessary and sufficient that*

1) *$F'(a_i)$ has at most one 1 in each column* $(i = 0, 1, \ldots, r-1)$.

2) *The boolean spectral radius of $S = F'(a_{r-1}) \ldots F'(a_0)$ is zero.*

a) The sufficient conditions. If 1) is satisfied then we show that $F(V_{a_i}) \subset V_{a_{i+1}}$ $(i = 1, 2, \ldots, r-1 \bmod r)$.

---

* Open question: How many distinct points a_i can be placed on the n-cube $\{0, 1\}^n$ (which contains 2^n points) such that the V_{a_i} are mutually disjoint? (in other words, compute the maximum length of an attractive cycle).

Indeed, if $y \in V_{a_i}$ then

$$d(F(y), \underbrace{F(a_i)}_{a_{i+1}}) = F'(a_i) \underbrace{d(y, a_i)}_{e_j}.$$

The columns of $F'(a_i)$ may only be zero or a basis vector according to the hypothesis. This shows that $F(y)$ belongs to the immediate neighbourhood of $a_{i+1} = F(a_i)$.

F therefore clearly maps V_{a_i} into $V_{a_{i+1}}$.

Suppose now (in order to simplify the notation and without restricting the generality) that x^0 is in V_{a_0}.

Then $x^s = F(x^{s-1})$ $(s = 1, 2, \ldots)$ belongs to V_{a_i}, with $i = s (\bmod r)$ and one has, according to the preceding, that

$$\begin{aligned} &x^0 \in V_{a_0} \\ &d(x^1, a_1) = F'(a_0)\, d(x^0, a_0) \qquad \text{and } x_1 \in V_{a_1} \\ &d(x^2, a_2) = F'(a_1)\, d(x^1, a_1) \qquad \text{and } x_2 \in V_{a_2} \\ &\vdots \\ &d(x^r, a_0) = F'(a_{r-1})\, d(x^{r-1}, a_{r-1}) \qquad \text{and } x_r \in V_{a_0} \end{aligned}$$

from which

$$d(x^r, a_0) = \underbrace{F'(a_{r-1})\, F'(a_{r-2}) \ldots F'(a_0)}_{S}\, d(x^0, a_0).$$

We then have, for all p, that

$$d(x^{rp}, a_0) = S^p\, d(x^0, a_0).$$

However, since S is strictly lower triangular up to a permutation (see Chap. 3, Theorem 5, p. 49) there exists a $p \le n$ such that $S^p = 0$ (boolean power) from which

$$x^{rp} = a_0.$$

The iterates $x^{rp+1}, x^{rp+2}, \ldots$ therefore run indefinitely through the cycle $a_1, a_2, \ldots, a_{r-1}, a_0$, which we were to show.

b) The necessary conditions. Suppose that the cycle C is attractive in its immediate neighbourhood. Assume now, however, that one of the derivative matrices, say $F'(a_0)$ (without restricting the generality) has more than one 1 in one of its columns. We might now easily show as in Theorem 4 that $F(V_{a_0})$ cannot be contained in V_{a_1}, because there then exists a point of V_{a_0} whose image is not in the immediate neighbourhood of a_1. This is, however, not possible according to the hypothesis.

For x^0 taken from V_{a_0} the successive iterates then run through the V_{a_j}. The calculations above are therefore valid and it follows that

$$d(x^{rp}, a_0) = S^p\, d(x^0, a_0) \qquad \text{for all } p.$$

By hypothesis, there exists $p \leq n$ such that we furthermore have $x^{rp} = a_0$ from which, clearly, $x^{rn} = a_0$, that is,

$$S^n d(x^0, a_0) = 0$$

independent of x^0 in V_{a_0}. Since $d(x^0, a_0)$ is a basis vector in $\{0, 1\}^n$ one concludes that all the columns of S^n are zero, which characterizes the fact that the boolean spectral radius of S is zero. □

Example. $n = 3$. We take the cycle $C = \{d, e\}$.

$$V_d = \{d, b, c, h\}$$
$$V_e = \{e, a, g, f\}.$$

(For $n = 3$, we observe that there may not be more than *two* points with disjoint immediate neighbourhoods.) Let F be defined by

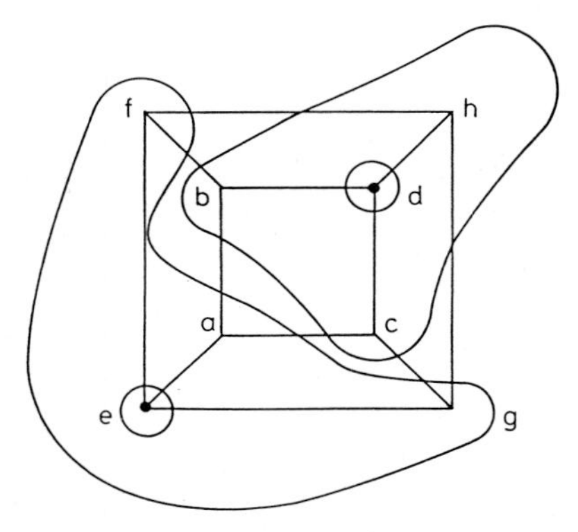

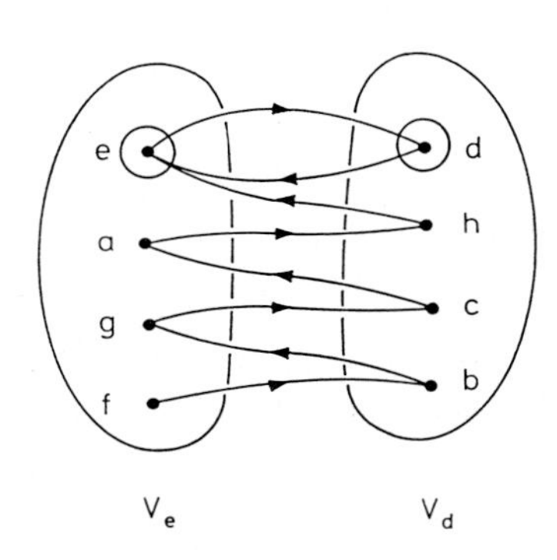

The cycle $C = \{d, e\}$ is then clearly attractive in its immediate neighbourhood. The table for F is given by

	x			$F(x)$			
a	0	0	0	1	1	1	h
b	0	0	1	1	1	0	g
c	0	1	0	0	0	0	a
d	0	1	1	1	0	0	e
e	1	0	0	0	1	1	d
f	1	0	1	0	0	1	b
g	1	1	0	0	1	0	c
h	1	1	1	1	0	0	e

and after some calculations we get

$$f_1(x) = \bar{x}_1 \bar{x}_2 + x_2 x_3$$
$$f_2(x) = \bar{x}_1 \bar{x}_2 + x_1 \bar{x}_3$$
$$f_3(x) = \bar{x}_2 \bar{x}_3 + x_1 \bar{x}_2.$$

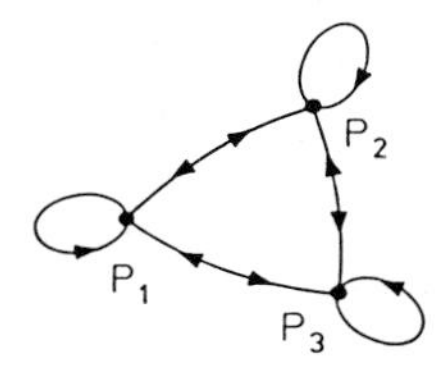

The connectivity graph for F is complete (the incidence matrix $B(F)$ is full of 1's).

We verify that

$$F'(d)=\begin{bmatrix}0&0&1\\0&1&0\\0&0&0\end{bmatrix} \quad \text{and} \quad F'(e)=\begin{bmatrix}1&0&0\\0&0&1\\0&1&0\end{bmatrix}$$

have no more than one 1 in each column. Furthermore

$$F'(d)\,F'(e)=\begin{bmatrix}0&1&0\\0&0&1\\0&0&0\end{bmatrix}$$

as well as

$$F'(e)\,F'(d)=\begin{bmatrix}0&0&1\\0&0&0\\0&1&0\end{bmatrix}$$

both have boolean spectral radius zero.

Moreover, if we write $H=F\circ F=F^2$, then we obtain

	x	$H(x)$	
a	0 0 0	1 0 0	e
b	0 0 1	0 1 0	c
c	0 1 0	1 1 1	h
d	0 1 1	0 1 1	d
e	1 0 0	1 0 0	e
f	1 0 1	1 1 0	g
g	1 1 0	0 0 0	a
h	1 1 1	0 1 1	d

where d and e are clearly fixed points of $H=F^2$. We furthermore have

$$H'(d)=\begin{bmatrix}0&0&1\\0&0&0\\0&1&0\end{bmatrix}=F'(e)\,F'(d)$$

$$H'(e)=\begin{bmatrix}0&1&0\\0&0&1\\0&0&0\end{bmatrix}=F'(d)\,F'(e).$$

One verifies that e and d are attractive fixed points (for H) in their immediate neighbourhoods.

Remarks. (1) If the cycle C is attractive in its immediate neighbourhood then it follows that we not only have

$$\rho(F'(a_{r-1}) \ldots F'(a_0)) = 0$$

but, clearly, also

$$\begin{aligned} &\rho(F'(a_0)\,F'(a_{r-1}) \ldots F'(a_1)) = 0 \\ &\rho(F'(a_1)\,F'(a_0) \quad \ldots F'(a_2)) = 0 \\ &\vdots \\ &\rho(F'(a_{r-2}) \qquad \ldots F'(a_0)\,F'(a_{r-1})) = 0. \end{aligned}$$

Since $\rho(AB) = \rho(BA)$ for any two $n \times n$ boolean matrices A and B.

(2) Moreover, for $r=1$ the cycle C reduces to a fixed point of F and Theorem 6 above simply restates the result of Theorem 4 (characterization of an attractive fixed point in its immediate neighbourhood).

(3) We might, in an analogous fashion, establish a generalization of Theorem 5 above by establishing a sufficient condition for the convergence of the iteration $x^{s+1} = F(x^s)$ "in a massive neighbourhood of a cycle". This is left as an exercise for the reader!

(4) *Once more, the notion of a cycle that is attractive in its immediate neighbourhood, may be easily interpreted in the context of an automata network.* Let $a_0, a_1, \ldots, a_{r-1}$ be r distinct configurations that are chained indefinitely in the parallel iteration for F (cycle $C = \{a_0, a_1, \ldots, a_{r-1}\}$).

If we then start from a configuration in the immediate neighbourhood of one of the a_i*'s* (that is to say, modifying one and only one cell in the configuration a_i) *and if the automata are made to operate in parallel, then we finish by returning to the cycle* $\{a_0, a_1, \ldots, a_{r-1}\}$ *while only passing through immediate neighbourhood configurations of the* a_j*'s.*

The preceding example is illustrated below.

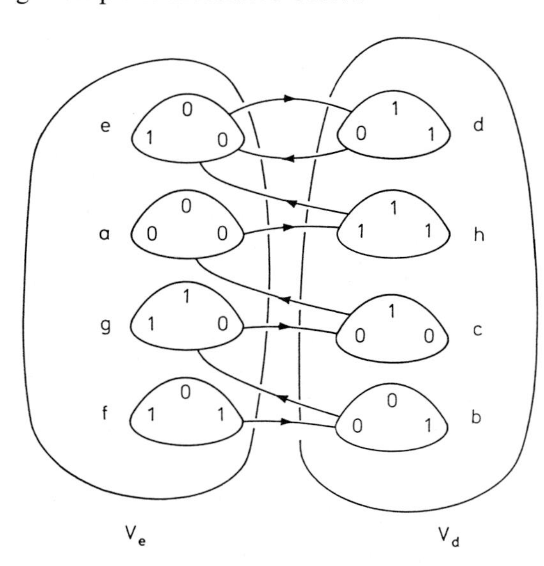

(5) The above study leads naturally to the establishment of the following result:

Theorem 7. *If* $a_0, a_1, \ldots, a_{r-1}$ *forms a cycle for* F *such that*

$$F(V_{a_0}) \subset V_{a_1};\ F(V_{a_1}) \subset V_{a_2};\ \ldots;\ F(V_{a_{r-1}}) \subset V_{a_0}$$

then setting

$$H = F^r \qquad (r \text{ fold composition})$$

one has

$$H'(a_0) = F'(a_{r-1}) \ldots F'(a_1)\, F'(a_0)$$

as well as analogous relations obtained by circular permutations.

This is a particular result for the derivative of an r-fold composition at an element of a cycle.

We confine ourselves to proving the result for $r=2$.

Then $F(a_0)=a_1$ $F(a_1)=a_0$ and for all j

$$d(\underbrace{F(a_0)}_{a_1}\, F(\tilde{a}_0^j)) = F'(a_0)\, \underbrace{d(a_0, \tilde{a}_0^j)}_{e_j}.$$

However, $\tilde{a}_0^j \in V_{a_0}$ and $F(V_{a_0}) \subset V_{a_1}$, which implies

$$F(\tilde{a}_0^j) \in V_{a_1}.$$

Let $b = F(\tilde{a}_0^j)$. Then, according to Theorem 2

$$d(F(a_1), F(b)) = F'(a_1)\, d(a_1, b),$$

from which one obtains, finally,

$$d(H(a_0), H(\tilde{a}_0^j)) = F'(a_1)\, F'(a_0)\, \underbrace{d(a_0\, \tilde{a}_0^j)}_{e_j}$$

and

$$d(H(a_0), H(\tilde{a}_0^j)) = \qquad H'(a_0)\, \overbrace{d(a_0\, \tilde{a}_0^j)}.$$

Since this is true for all j between 1 and n one has, necessarily

$$H'(a_0) = F'(a_1)\, F'(a_0).$$

Similarly one shows that

$$H'(a_1) = F'(a_0)\, F'(a_1).$$ □

We refer to the last example for an illustration of this result.

We can also clearly see how this proof may be generalized to an arbitrary r as well as see the role played by the inclusion $F(V_{a_0}) \subset V_{a_1}$.

The following is a counterexample showing what happens when the above condition is not satisfied.

$$n=3 \qquad F(a)=h \qquad F(h)=a.$$

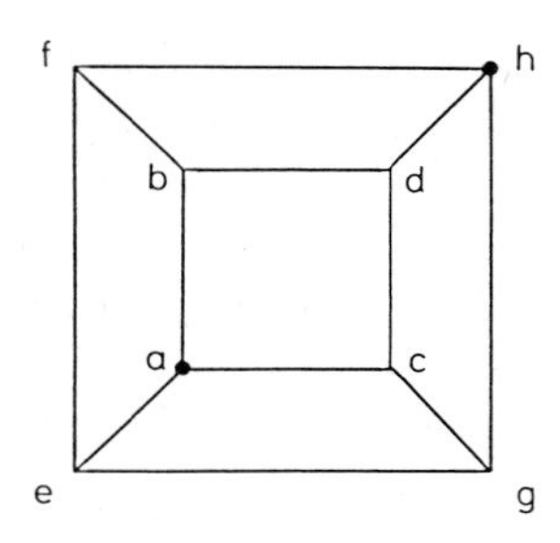

The example is devised so that $F(b)=a$ and that b is in the neighbourhood of a. $F(b)$, however, is not in the neighbourhood of $F(a)=h$.

	x	$F(x)$	$F^2(x)$ $=H(x)$
a	0 0 0	1 1 1	0 0 0
b	0 0 1	0 0 0	1 1 1
c	0 1 0	1 0 0	1 1 0
d	0 1 1	1 1 0	1 1 0
e	1 0 0	1 1 0	1 1 0
f	1 0 1	0 1 1	1 1 0
g	1 1 0	1 1 0	1 1 0
h	1 1 1	0 0 0	1 1 1

Then

$$F'(a)=\begin{bmatrix}0&0&1\\0&1&1\\1&1&1\end{bmatrix} \qquad F'(h)=\begin{bmatrix}1&0&1\\1&1&1\\0&1&0\end{bmatrix}$$

$$H'(a)=\begin{bmatrix}1&1&1\\1&1&1\\0&0&1\end{bmatrix} \neq F'(h)\,F'(a)=\begin{bmatrix}1&1&1\\1&1&1\\0&1&1\end{bmatrix}$$

$$H'(h)=\begin{bmatrix}0&0&0\\0&0&0\\1&1&1\end{bmatrix} \neq F'(a)\,F'(h)=\begin{bmatrix}0&1&0\\1&1&1\\1&1&1\end{bmatrix}.$$

(6) *Last remark.* We indicated above that if a cycle $C=\{a_0, a_1, \ldots, a_{r-1}\}$ is attractive in its immediate neighbourhood then the fixed points $a_0, a_1, \ldots, a_{r-1}$ of F^r are necessarily attractive in their immediate neighbourhoods.

We verify this for a_0 for example.

From Theorem 6, all the $F'(a_i)$ have at most one 1 per column, This is therefore also true of their product $S = F'(a_{r-1}) \dots F'(a_0)$. We know, however, that, according to Theorem 6, the spectral radius $\rho(S)$ of S is zero.

Now, from the preceding Theorem 7, we have

$$S = H'(a_0) \quad \text{and} \quad H(x) = F^r(x).$$

One has, therefore, rediscovered the two conditions guaranteeing that a_0 is attractive (for H) in its immediate neighbourhood (see Theorem 4):

1) $H'(a_0)$ has no more than a single 1 per column.

2) $\rho(H'(a_0) = 0$.

One rediscovers also that since all the a_i's are attractive fixed points (for F^r) in their immediate neighbourhoods, then the neighbourhoods V_{a_i} are necessarily pairwise disjoint.

9. Conclusions

The notion of the *discrete derivative* introduced in this chapter allows us to deepen the *local study of iterations on* $\{0,1\}^n$. This discrete derivative is connected to the incidence matrix of the operator studied (Theorem 1) and also to the topological tool d used (Theorem 2).

Thanks to this tool one may characterize the notion of an *attractive fixed point* (Theorem 4) and of an *attractive cycle* (Theorem 6) in their immediate neighbourhoods. One has also given sufficient conditions such that a fixed point should be attractive in a *massive neighbourhood* (Theorem 5) thanks to a formula "of finite increase" (Theorem 3) valid in the discrete case.

[However, one may not take the derivative of a composition of functions in the usual manner, except for a very special case (Theorem 7).]

Thus, the analogy has been "carried rather far" and the results obtained may be applied directly to the study of the local evolution of a network of automata.

We also mention that we restricted (for reasons of simplicity) the configuration space to $\{0,1\}^n$ (each cell of the network may only take on the states 0 or 1). The study for the general case where $X = \prod_{i=1}^{n} X_i$ (X_i finite) results in some more complicated definitions as well as more complicated usages of the discrete derivative (see [79]).

7. A Discrete Newton Method

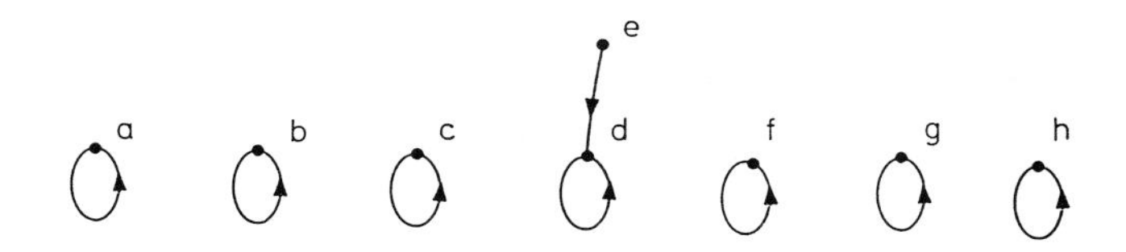

We will now conclude our sight-seeing tour of the behaviour of discrete iterations by defining and analyzing *a discrete Newton method*. The usual Newton method (for a mapping F of $\mathbb{R}^n \to \mathbb{R}^n$) is based on the concept of a derivative. Since we introduced a *discrete* derivative in the previous chapter, it seems very natural to try to carry over the ideas behind Newton's method in the continuous setting into the discrete context.

Therefore, let F be a map of $\{0, 1\}^n$ into itself. This map was interpreted in terms of automata networks in Chap. 1. Here we visualize F in a different manner. We consider the interpretation of F in terms of a "black box" having n binary inputs and n binary outputs that are functions of the inputs. Schematically this may be visualized as

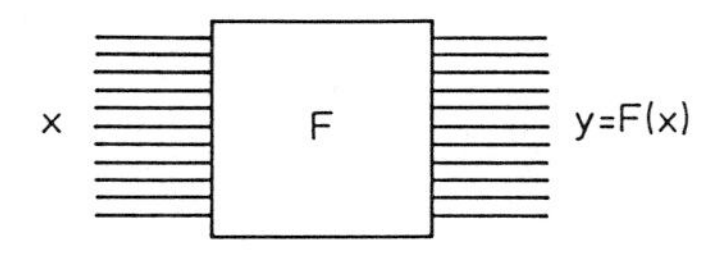

Clearly, if a boolean vector x of n entries is entered into the box then the output from the box is a boolean vector $y = F(x)$.

We wish to find a solution to the following problem:

"How do we know if there exists an $x \in \{0, 1\}^n$ satisfying $F(x) = 0$ (the zero vector)? If the question is answered affirmatively, calculate such an x."

The first method one may think of here is clearly the method of exhaustion. This simply consists of trying each of the 2^n possible vectors x. For each vector x that is tested the cost is one evaluation of F and between 1 and n comparisons of the components with 0. The cost of the exhaustive method is, therefore

$$\begin{cases} 2^n \text{ evaluations of } F \\ \text{between } 2^n \text{ and } n\,2^n \text{ comparisons of the components with } 0. \end{cases}$$

This is, of course, very expensive. It does, however, give a complete description of all the roots of F, and if there are none the result is "no roots".

Can one proceed in a different manner? Without making further assumptions on F (monotonicity and so on), one does not know how to proceed except by repeated trials (see [162]).

We will therefore develop a *Newton method* in a discrete setting in analogy with the Newton method in numerical analysis used to find solutions of non-linear equations $F(x)=0$ (where F maps $\mathbb{R}^n$ into itself). The method we develop is an *iterative method*. Starting from an arbitrary x^0 (that is to say, one of the 2^n vertices of the n-cube), one computes a sequence of boolean vectors $x^1, x^2, x^3, \ldots, x^r, \ldots$. The hope is that the n-th vector x^n will be a root of F *(stationary convergence in at most n steps towards a root)*. Similar to the usual Newton method, it is not possible to discuss this convergence exhaustively and to give all the roots. One is indeed limited (if the algorithm converges, which is not always assured*) to give *one* root of F.

Clearly, the problem of finding a z such that $F(z)=\alpha$ (α given, not necessarily zero, in $\{0,1\}^n$) is treated in the same manner, without further complications.

1. Context

We work in the following context. The basic space $\{0,1\}^n$ (the 2^n vertices of the n-cube) is here identified with $(Z/2)^n$ and we use in particular the addition $\oplus$ in $(Z/2)^n$, remembering that for all x in $(Z/2)^n$ we have $x \oplus x = 0$ (the zero vector).

For all x in $(Z/2)^n$ the *immediate neighbourhood* V_x may then be described by

$$V_x = \{x, x \oplus e_1, \ldots, x \oplus e_n\}$$

where the e_j are the basis vectors in $(Z/2)^n$. Clearly, one has

$$\tilde{x}^j = x \oplus e_j \qquad (j\text{-th neighbour of } x).$$

* The most important practical question about the algorithm that we define is its efficiency, that is, whether the algorithm converges both often and fast (in the same manner as the usual Newton method). This will be elaborated in the sequel.

The topological tool used in $(Z/2)^n$ is once more the boolean vector distance d, which may be expressed as

$$\forall x, y \in (Z/2)^n \qquad d(x, y) = x \oplus y.$$

Beware: For all x and y in $(Z/2)^n$, $d(x, y)$ must be considered to be a *boolean* vector having n components. In particular, in the triangle inequality

$$\forall x, y, z \in (Z/2)^n \qquad d(x, z) \leq d(x, y) + d(y, z)$$

(componentwise inequality) the $+$ is a *boolean addition* of the vectors (corresponding to $1+1=1$) and not the $\oplus$ addition in $(Z/2)^n$.

One recalls that the (stationary) convergence of a sequence x^r from $(Z/2)^n$ towards a limit x in fact means that after a finite number of steps one has $d(x^r, x) = 0$, that is, $x^r = x$.

F is then a map of $(Z/2)^n$ into itself having components f_i. One wishes to find a root of F. The relation $y = F(x)$ may also be written

$$y_i = f_i(x_1, \ldots, x_n) \qquad (i = 1, 2, \ldots, n).$$

$F'(x)$ is the discrete derivative of F at some x in $\{0, 1\}^n$. It is an $n \times n$ matrix with elements from $\{0, 1\}$. These elements are obtained in the following manner (see also the preceding chapter): $F'(x)$ has a 1 in position (i, j) if $f_i(x)$ is different from $f_i(\tilde{x}^j)$, otherwise 0.

The discrete derivative and the vector distance d are connected by the following formula:

For all x in $(Z/2)^n$ and all immediate neighbours y to x $(y \in V_x)$ one has

$$d(F(x), F(y)) = F'(x) \otimes d(x, y) = F'(y) \otimes d(x, y).$$

Indeed, y being a neighbour for x may be written $y = x \oplus e_j = \tilde{x}^j$. Then $d(x, y) = x \oplus y = e_j$ and $d(F(x), F(y))$ is nothing but the j-th column of the matrix $F'(x)$ [which is also evidently, by construction, that of $F'(y)$]. One notes that since $d(x, y)$ is the basis vector e_j, the multiplication $\otimes$ that is used may be either the multiplication "in $Z/2$" or the boolean multiplication (used in the preceding chapter).

We now seek a root of F. By analogy with numerical analysis we therefore study the following procedure:

$$\begin{cases} x^0 \text{ arbitrary in } (Z/2)^n \\ x^{r+1} = x^r \oplus [A_r^{-1} \otimes F(x^r)] \qquad (r = 0, 1, 2, \ldots) \end{cases}$$

where A_r $(r = 0, 1, 2, \ldots)$ is a sequence of matrices with elements in $Z/2$ that furthermore are *invertible as operators on* $(Z/2)^n$.

$\oplus$ denotes the addition in $(Z/2)^n$ and $\otimes$ represents the matrix multiplication "in $(Z/2)^n$".

Then the choice $A_r = F'(x^r)$ (assumed invertible) gives the *standard Newton* method. Similarly, the choice $A_r = A$ (a constant invertible matrix) gives the *simplified Newton* method.

Practically the algorithm clearly proceeds as follows:

$$\begin{cases} x^r \text{ known, calculate } F(x^r) \text{ and } A_r \\ \text{solve the system } A_r \otimes z = F(x^r) \text{ (in } (Z/2)^n\text{, assumed solvable)} \\ \text{calculate } x^{r+1} = x^r \oplus z. \end{cases}$$

It is clear that the algorithm is stationary $(z=0)$ starting from the moment where x^r is a root of F $(F(x^r)=0)$.

Each step of the standard Newton method $(A_r = F'(x^r))$ costs:

Evaluation of $F'(x^r)$ $\begin{cases} \text{requires the evaluation of } F \text{ at } x^r \text{ and} \\ \text{its } n \text{ neighbours for a total of } n+1 \\ \text{evaluations of } F. \text{ In addition there} \\ \text{are } n^2 \text{ comparisons of numbers.} \end{cases}$

Solving the linear system $\begin{cases} \text{Order of } \dfrac{2n^3}{3} \text{ additions and} \\ \text{multiplications in } Z/2 \text{ (Gauss)} \end{cases}$

$x^{r+1} = x^r \oplus z$ n additions in $Z/2$.

Since one hopes to succeed in no more than n steps, the cost for the standard Newton algorithm is estimated to only be the order of

$$\begin{cases} n(n+1) \text{ evaluations of } F \\ n^3 \text{ comparisons of numbers} \\ \dfrac{2n^4}{3} \text{ additions and multiplications in } Z/2. \end{cases}$$

This cost, clearly lower than the exhaustive method, may suggest a practical method for finding a root of F. *In the same manner as in the usual Newton method, the power of the method is due to the fact that local information for F is used, provided by the knowledge of $F'(x^r)$.*

Finally, if the task is to solve $F(x) = \alpha$ (α a given vector in $(Z/2)^n$), the standard Newton algorithm is written as

$$\begin{cases} x^r \text{ known. Calculate } F(x^r) \text{ and } F'(x^r). \\ \text{Solve } F'(x^r) \otimes z = F(x^r) \oplus \alpha \\ x^{r+1} = x^r \oplus z. \end{cases}$$

2. Two Simple Examples

Example 1. $n=3$, F is given by the table

	x			$F(x)$		
a	0	0	0	1	0	0
b	0	0	1	1	1	0
c	0	1	0	1	0	1
d	0	1	1	0	0	0
e	1	0	0	0	0	1
f	1	0	1	0	1	1
g	1	1	0	1	1	0
h	1	1	1	1	1	1

There is one root $d=\begin{bmatrix}0\\1\\1\end{bmatrix}$. Take for example $x^0=h=\begin{bmatrix}1\\1\\1\end{bmatrix}$. $F(x^0)=\begin{bmatrix}1\\1\\1\end{bmatrix}$

and $F'(x^0)=\begin{bmatrix}1&1&0\\1&0&0\\1&0&1\end{bmatrix}$. Solving $F'(x^0)\otimes z=F(x^0)$ we get $z=\begin{bmatrix}1\\0\\0\end{bmatrix}$ and

$x^1=x^0\oplus z=\begin{bmatrix}0\\1\\1\end{bmatrix}$. *This is the solution.*

If one starts again with different initial vectors one may display the *standard Newton iteration graph*

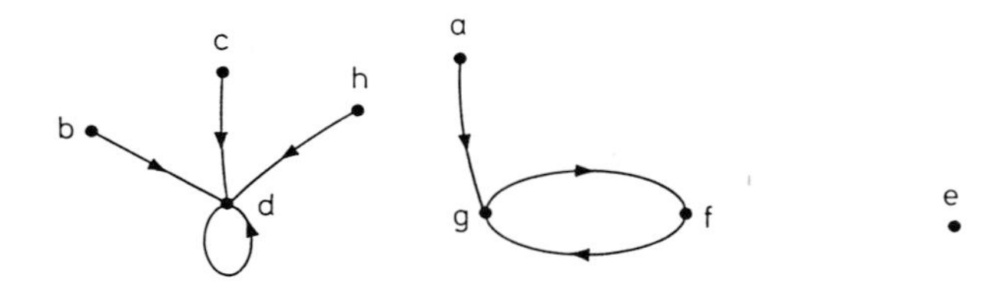

This example allows us to illustrate all the possible cases.

With some starting vectors (here b, c, h) one obtains a solution (here in one step).

Certain other vectors may produce a cyclical behaviour (here a, g, f).

Finally, for e the algorithm is not defined ($F'(e)$ is singular).

In the following example one obtains many singular systems. Nevertheless, wherever possible, one still solves the system (and therefore obtains several solutions).

Example 2. $n=3$, F is given by the table

	(x)	$F(x)$
a	0 0 0	1 0 0
b	0 0 1	0 0 0
c	0 1 0	1 1 1
d	0 1 1	0 1 0
e	1 0 0	0 0 0
f	1 0 1	1 0 0
g	1 1 0	1 1 0
h	1 1 1	0 1 0

There are two roots, b and e. Take, for example, $x^0 = \begin{bmatrix} 0 \\ 0 \\ 0 \end{bmatrix} = a$.

Then $F(x^0) = \begin{bmatrix} 1 \\ 0 \\ 0 \end{bmatrix}$ and $F'(x^0) = \begin{bmatrix} 1 & 0 & 1 \\ 0 & 1 & 0 \\ 0 & 1 & 0 \end{bmatrix}$ is singular.

But the system $F'(x^0) \otimes z = F(x^0)$ is not impossible. One solves it and one obtains two solutions

$$z_1 = \begin{bmatrix} 1 \\ 0 \\ 0 \end{bmatrix} \quad \text{and} \quad z_2 = \begin{bmatrix} 0 \\ 0 \\ 1 \end{bmatrix}.$$

From this one has two possibilities for x^1:

$$x^0 \oplus z_1 = \begin{bmatrix} 1 \\ 0 \\ 0 \end{bmatrix} = e. \qquad \textit{This is a root.}$$

$$x^0 \oplus z_2 = \begin{bmatrix} 0 \\ 0 \\ 1 \end{bmatrix} = b. \qquad \textit{This is the other root.}$$

Here, we have the complete iteration graph for the standard Newton method for this example

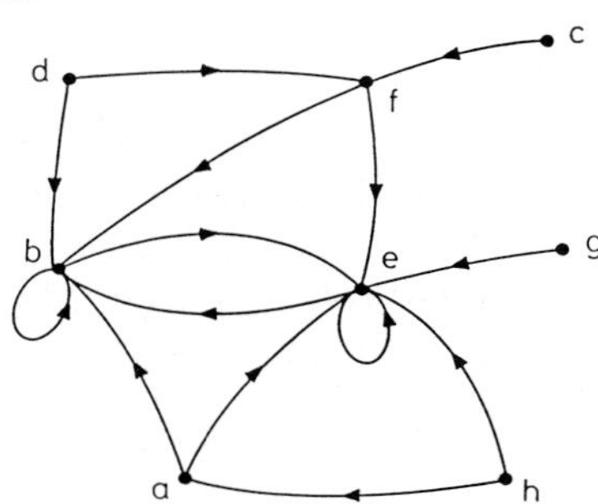

(The starting points c and g are the only ones not generating a singular system.)

We may therefore convince ourselves in an intuitive manner that the standard Newton method, applied to some fairly small examples, displays a certain efficiency in searching for a root of F.

3. Interpretation in Terms of Automata

Let us consider the automata network associated with F, with the interpretation given in the first chapter:

To search for an x in $\{0,1\}^n$ such that $F(x)=0$, (or more generally $F(x)=\alpha$ for a given α in $\{0,1\}^n$) *that is, to search whether the configuration* 0 *(respectively α) has an antecedent in the parallel iteration for F.*

The problem is simply to find whether a given configuration has a configuration generating it. This is a basic problem of automata networks.

Example. $n=3$. Consider the network

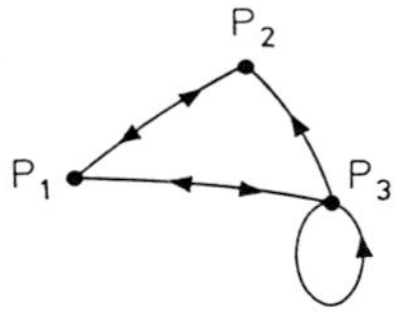

P_1 P_2 P_3

from which $B(F)=\begin{bmatrix} 0 & 1 & 1 \\ 1 & 0 & 1 \\ 1 & 0 & 1 \end{bmatrix}$.

Let us for example take

$$f_1(x)=x_2+x_3$$
$$f_2(x)=x_1 x_3$$
$$f_3(x)=x_1+x_3 \quad \text{(boolean notation).}$$

The table for F is therefore

	x	$F(x)$	
a	0 0 0	0 0 0	a
b	0 0 1	1 0 1	f
c	0 1 0	1 0 0	e
d	0 1 1	1 0 1	f
e	1 0 0	0 0 1	b
f	1 0 1	1 1 1	h
g	1 1 0	1 0 1	f
h	1 1 1	1 1 1	h

There is one root which is a.

We now show the iteration graph for F, and, after some calculations, the iteration graph for the standard Newton method.

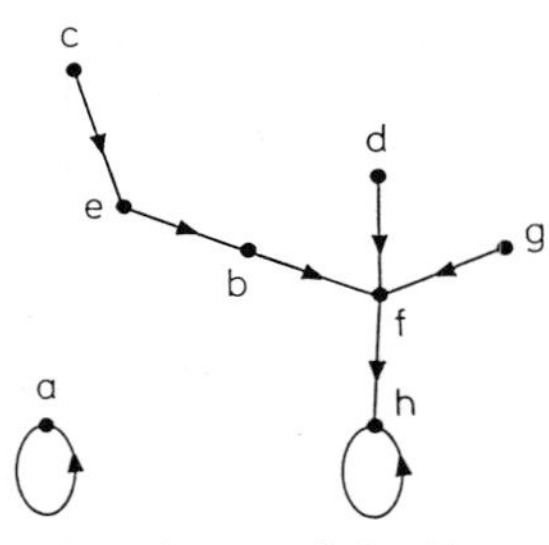

Iteration graph for F

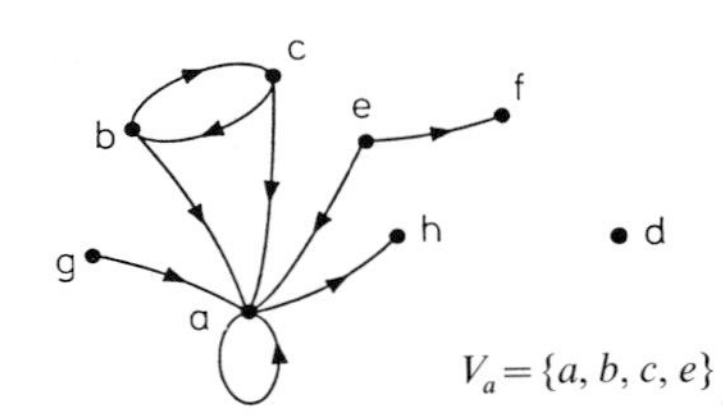

Iteration graph for the standard Newton method for F

4. The Study of Convergence: The Case of the Simplified Newton Method

We will start our convergence studies with the simplified Newton method for the sake of simplicity where

$$x^{r+1} = x^r \oplus [A^{-1} \otimes F(x^r)] \quad (r = 0, 1, 2, \ldots)$$

and where A is chosen invertible. This is a method of successive approximations

$$x^{r+1} = H(x^r) \quad (r = 0, 1, 2, \ldots)$$

where the operator H is simple enough to be easily described by

$$\forall x \in (Z/2)^n \quad H(x) = x \oplus [A^{-1} \otimes F(x)]$$

and where we get

$$\forall x \in (Z/2)^n \quad H'(x) = I \oplus [A^{-1} \otimes F'(x)].$$

A) Local Convergence in the Immediate Neighbourhood of a Root of F

By using H' one may apply the *local convergence results* established in the preceding chapter. *These results will guarantee the convergence to a root in no more than n steps.*

Theorem 1. *Let a be a root of F (that is to say, a fixed point of H).*

In order that a will be attractive in its immediate neighbourhood it is necessary and sufficient that

1) $I \oplus [A^{-1} \otimes F'(a)]$ has at most one 1 per column.

2) The boolean spectral radius of $I \oplus [A^{-1} \otimes F'(a)]$ is zero (that is to say, there exists a permutation matrix P such that $P^t[I \oplus [A^{-1} \otimes F'(a)]]P$ is strictly lower triangular).

We recall (see preceding chapter, Theorem 4, p. 103) that 1) assures that the sequence x^r remains in V_a for all x^0 taken from V_a and that 2) assures the *convergence to the root a in no more than n steps.* □

Remark. Conditions 1) and 2) express the fact that "A should not be very different from $F'(a)$". Moreover, the choice of $A = F'(a)$ (assumed invertible) satisfies conditions 1) and 2) since the matrix $I \oplus [A^{-1} \otimes F'(a)]$ is then the zero matrix.

We then in fact have $x^1 = a$ for any x^0 taken from V_a (the immediate neighbourhood of the root a of F). Indeed, since $x^0 \in V_a$ it follows that x^0 and a are neighbours and one has

$$F'(a) \otimes d(x^0, a) = d(F(x^0), \underbrace{F(a)}_{0}) = F(x^0).$$

Thus

$$\begin{aligned} x^1 &= x^0 \oplus [F'(a)]^{-1} \otimes F(x^0) = x^0 \oplus \underbrace{d(x^0, a)} \\ &= x^0 \oplus x^0 \oplus a = a. \end{aligned}$$

From this we have

Corollary. *Let a be a root of F and let $x^0 \in V_a$. Then for the choice $A = F'(a)$ (assumed invertible) one has $x^1 = a$.*

Example. $n = 3$, F is given by the table

	x	$F(x)$
$\tilde{a}^2$	0 0 0	1 1 1
	0 0 1	0 1 0
$\tilde{a}^3$	0 1 0	1 0 0
a	0 1 1	0 0 0
	1 0 0	0 0 1
	1 0 1	0 1 1
	1 1 0	1 1 1
$\tilde{a}^1$	1 1 1	0 0 1

There is one root $a = (0\ 1\ 1)$ and

$$F'(a) = \begin{bmatrix} 0 & 0 & 1 \\ 0 & 1 & 0 \\ 1 & 0 & 0 \end{bmatrix} = [F'(a)]^{-1}.$$

1) We choose $A=\begin{bmatrix}0&1&1\\0&1&0\\1&0&0\end{bmatrix}$ from which $A^{-1}=\begin{bmatrix}0&0&1\\0&1&0\\1&1&0\end{bmatrix}$ and

$$I\oplus[A^{-1}\otimes F'(a)]=\begin{bmatrix}0&0&0\\0&0&0\\0&1&0\end{bmatrix}.$$

Conditions 1) and 2) of Theorem 1 are satisfied. In fact

– For $x^0=\tilde{a}^1=\begin{bmatrix}1\\1\\1\end{bmatrix}$ we get $F(x^0)=\begin{bmatrix}0\\0\\1\end{bmatrix}$ from which

$$x^1=\begin{bmatrix}1\\1\\1\end{bmatrix}\oplus\begin{bmatrix}0&0&1\\0&1&0\\1&1&0\end{bmatrix}\begin{bmatrix}0\\0\\1\end{bmatrix}=\begin{bmatrix}0\\1\\1\end{bmatrix}=a.$$

Analogously for $x^0=\tilde{a}^2$ we get $x^1=\tilde{a}^3$
for $x^0=\tilde{a}^3$ we get $x^1=a$.

We show a part of the iteration graph for the simplified Newton method below $\left(\text{for the choice } A=\begin{bmatrix}0&1&1\\0&1&0\\1&0&0\end{bmatrix}\right)$

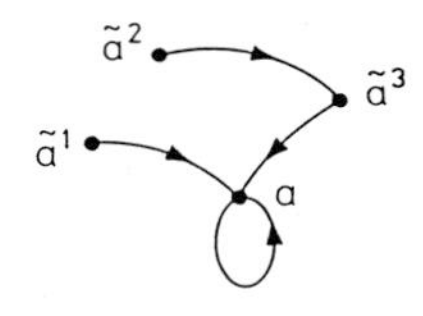

One verifies that a is a root of F that is attractive in its immediate neighbourhood.

2) For the same F, let us now take $A=F'(a)=\begin{bmatrix}0&0&1\\0&1&0\\1&0&0\end{bmatrix}=[F'(a)]^{-1}$

– for $x^0=\begin{bmatrix}1\\1\\1\end{bmatrix}=\tilde{a}^1,\qquad x^1=\begin{bmatrix}1\\1\\1\end{bmatrix}\oplus\begin{bmatrix}1\\0\\0\end{bmatrix}=\begin{bmatrix}0\\1\\1\end{bmatrix}=a$

– for $x^0=\begin{bmatrix}0\\0\\1\end{bmatrix}=\tilde{a}^2,\qquad x^1=\begin{bmatrix}0\\0\\1\end{bmatrix}\oplus\begin{bmatrix}0\\1\\0\end{bmatrix}=\begin{bmatrix}0\\1\\1\end{bmatrix}=a$

– for $x^0=\begin{bmatrix}0\\1\\0\end{bmatrix}=\tilde{a}^3$, $\quad x^1=\begin{bmatrix}0\\1\\0\end{bmatrix}\oplus\begin{bmatrix}0\\0\\1\end{bmatrix}=\begin{bmatrix}0\\1\\1\end{bmatrix}=a.$

One thus verifies (see the corollary) that for all $x^0\in V_a$, $x^1=a$.

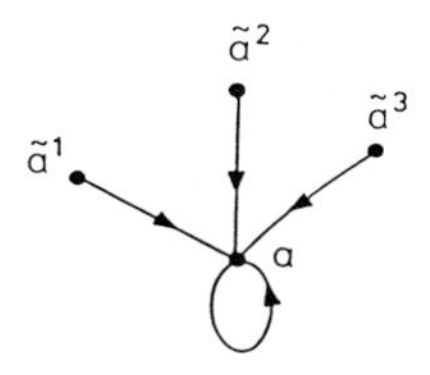

3) *Counterexample.* Let us choose

$$A=\begin{bmatrix}1&1&0\\0&1&1\\0&0&1\end{bmatrix}$$

from which

$$A^{-1}=\begin{bmatrix}1&1&1\\0&1&1\\0&0&1\end{bmatrix}$$

so that

$$I\oplus[A^{-1}\otimes F'(a)]=\begin{bmatrix}0&1&1\\1&0&0\\1&0&1\end{bmatrix}.$$

Neither of the conditions 1) or 2) of Theorem 1 are satisfied. After doing the calculations we get the following iteration graph:

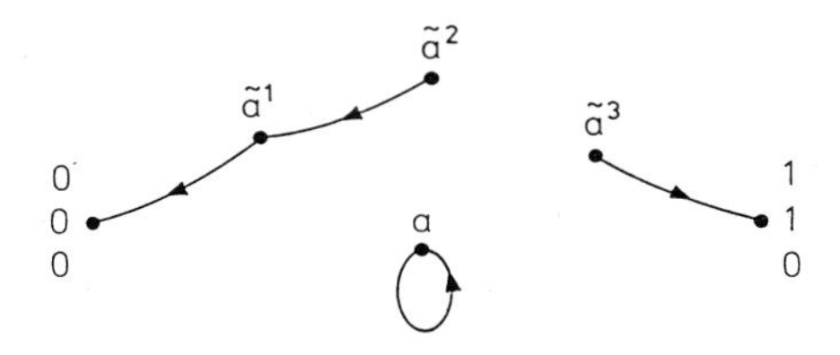

This clearly verifies that a is not a root that is attractive in its immediate neighbourhood.

B) Local Convergence in a Massive Neighbourhood of a Root of F

We will now similarly establish a result on the local convergence in *a massive* neighbourhood of a root a of F by using the results of the previous chapter (see Theorem 5, p. 110).

Theorem 2. *Let a be a root of F and let W_a be a massive neighbourhood of a. Let W_0 be the corresponding massive neighbourhood of* 0, *and define*

$$M = \operatorname*{Sup}_{z \in W_a} H'(z)$$

(*where the* Sup *is elementwise, with* $H'(z) = I \oplus [A^{-1} \otimes F'(z)]$). *In order that a shall be attractive in W_a it suffices that:*

1) $MW_0 \subset W_0$.

2) The boolean spectral radius of M is zero (there exists a permutation matrix P such that $P^t MP$ is strictly lower triangular).

(Recall that for x^0 from W_a, *the simplified Newton iteration remains in W_a and becomes stationary at a in at most n steps.*)

Example. Let F be taken as an affine map in $(Z/2)^n$

$$F(x) = [T \otimes x] \oplus u$$

where T is an $n \times n$ matrix with elements in $Z/2$ and where u is a given vector in $(Z/2)^n$ such that for all x in $(Z/2)$ we have

$$F'(x) = T.$$

Let us also choose T invertible. Then the equation $F(x) = 0$ has a single root $a = T^{-1} \otimes u$. Furthermore $H'(z) = I \oplus [A^{-1} \otimes T]$ is independent of z and the operator H is also affine and given by

$$H(x) = [[I \oplus [A^{-1} \otimes T]] \otimes x] \oplus [A^{-1} \otimes u]$$

such that, in this case,

$$M = I \oplus [A^{-1} \otimes T].$$

It is then sufficient to choose W_0 and A in order to satisfy conditions 1) and 2).

We take for $n = 3$

$$T = \begin{bmatrix} 0 & 1 & 0 \\ 1 & 1 & 0 \\ 1 & 1 & 1 \end{bmatrix}, \quad u = \begin{bmatrix} 1 \\ 0 \\ 1 \end{bmatrix}, \quad T^{-1} = \begin{bmatrix} 1 & 1 & 0 \\ 1 & 0 & 0 \\ 0 & 1 & 1 \end{bmatrix}.$$

The table for F is then the following:

	x			$F(x)$		
h	0	0	0	1	0	1
g	0	0	1	1	0	0
f	0	1	0	0	1	0
e	0	1	1	0	1	1
d	1	0	0	1	1	0
c	1	0	1	1	1	1
b	1	1	0	0	0	1
a	1	1	1	0	0	0

The root is

$$a=\begin{bmatrix}1\\1\\1\end{bmatrix}=T^{-1}\otimes u.$$

Let us take

$$A=\begin{bmatrix}0&1&0\\1&0&0\\0&0&1\end{bmatrix}=A^{-1}$$

and we get

$$M=I\oplus[A^{-1}\otimes T]=\begin{bmatrix}0&1&0\\0&0&0\\1&1&0\end{bmatrix}.$$

One may clearly take all of $(Z/2)^3$ as a massive neighbourhood of a, that is, $W_a=W_0=(Z/2)^3$. The conditions 1) and 2) are then verified. After some calculations we obtain the iteration graph for the simplified Newton method as

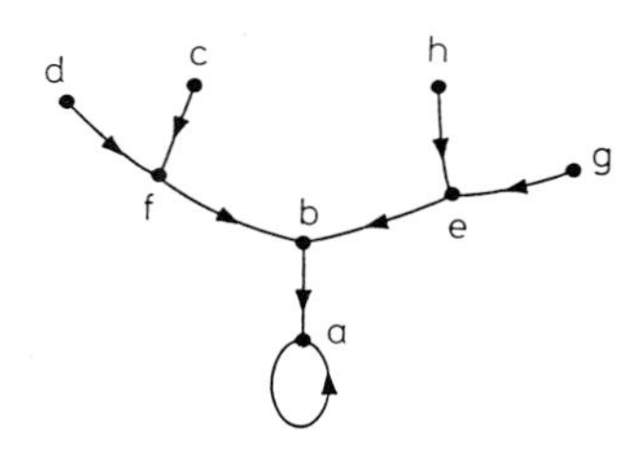

We verify that a is an attractive root for all of $(Z/2)^3$ and that this root is reached in no more than three steps in all cases (the iteration graph for the Newton method is simple).

Remark 1. Now let W_0 be the set of vectors in $(Z/2)^3$ having the second component zero.

$$W_0=\left\{\begin{bmatrix}0\\0\\0\end{bmatrix}\begin{bmatrix}1\\0\\0\end{bmatrix}\begin{bmatrix}0\\0\\1\end{bmatrix}\begin{bmatrix}1\\0\\1\end{bmatrix}\right\}=\{h,d,g,c\}.$$

Then

$$W_a=a\oplus W_0=\left\{\begin{bmatrix}1\\1\\1\end{bmatrix}\begin{bmatrix}0\\1\\1\end{bmatrix}\begin{bmatrix}1\\1\\0\end{bmatrix}\begin{bmatrix}0\\1\\0\end{bmatrix}\right\}=\{a,e,b,f\}$$

and W_a is a massive neighbourhood of the root a associated with W_0.

Condition 1) is once more satisfied. Since 2) is also satisfied, we are assured, according to Theorem 2, that a is attractive in all of W_a. This is verified and we have the following subgraph of the preceding graph:

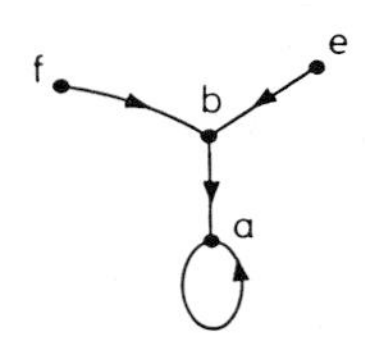

Remark 2. In the general case with F affine ($F(x)=[T\otimes x]\oplus u$, with T regular and $a=T^{-1}\otimes u$) take all of $(Z/2)^n$ for W_a and then also for W_0. Once more let A be chosen such that $M=I\oplus[A^{-1}\otimes T]$ has the boolean spectral radius zero. Then the conditions 1) and 2) are satisfied and a is attractive in all of $(Z/2)^n$. This is the case illustrated in the preceding example.

We will show that in this case the operator H is in fact contracting on $(Z/2)^n$ relative to d. Clearly one has

$$H(x)=\{I\oplus[A^{-1}\otimes F]\}(x)$$

from which

$$H(x)\oplus H(y)=\underbrace{(I\oplus[A^{-1}\otimes T])}_{M}\otimes(x\oplus y).$$

From this we obtain, noting by $\cdot$ the boolean matrix multiplication (*be careful*):

$$d(H(x),H(y))\leq M\cdot d(x,y)\qquad\forall x,y\in(Z/2)^n.$$

This therefore means that H is contracting on $(Z/2)^n$ (M has spectral radius zero). The iteration is therefore necessarily convergent independent of the starting vector x^0 in no more than n steps to the unique solution $T^{-1}\otimes u$ according to the basic result on contraction given in Chap. 4.

In this particular case we therefore rediscover the results of Theorem 2 above.

5. The Study of Convergence, the General Case

Let us now turn to the study of the iteration in the general case where

$$x^{r+1}=x^r \oplus [A_r^{-1} \otimes F(x^r)].$$

As with the usual method in $\mathbb{R}^n$, the study of the general case is more difficult. Let us therefore first remark that the standard Newton method will still deliver the solution in one step independent of x^0 for an affine F ($F(x) = [T \otimes x] \oplus u$ with T regular) since

$$\forall x \quad F'(x)=T \quad \text{and hence} \quad x^1=x^0 \oplus T^{-1} \otimes ([T \otimes x^0] \oplus u)=T^{-1} \otimes u$$

which is the unique solution of $[T \otimes x] \oplus u=0$.

It would be interesting to try to transpose the analysis of Kantorovich (for Newton's method in a Banach space) into our discrete context in order to study the general case. Furthermore, it would be useful to seek sufficient conditions for F guaranteeing the existence of a root for F (via the convergence of the iterative procedure studied towards this root). One might, in particular, seek a "discrete analog" to the following theorem valid in a Banach space.

Theorem (Kantorovich) [32]. *Let F be a Fréchet differentiable map of the Banach space E (with norm $\|\cdot\|$) into itself.*

Furthermore, let A_r be a sequence of invertible linear operators mapping E into E such that

$$\forall r \quad \|A_r^{-1}\| \leq M.$$

If there exists a ball B centered on x^0 in E with radius r

$$B=\{x \in E \; \|x-x^0\| \leq r\}$$

such that:

(i) $\beta=m \operatorname{Sup}_r \operatorname{Sup}_{x \in B} \|F'(x)-A_r\| < 1,$

(ii) $\|F(x^0)\| \leq \frac{r}{M}(1-\beta).$

Then the sequence

$$x^{r+1}=x^r-A_r^{-1} F(x^r) \qquad (r=0, 1, 2, \ldots)$$

remains in B and converges to a root a of F unique in B.

The convergence is furthermore geometric

$$\|x^r-a\| \leq c\, \beta^r.$$

The problem now consists of transposing the metric analysis contained in the proof of this theorem into the discrete setting. *This problem is still open!*

On the other hand, we may establish an explicit error formula. This formula will allow us to state a (complicated) necessary and sufficient condition for the "monotone progression" of the algorithm towards a root a (which is a priori assumed to exist).

Definition. We say that a sequence $x^0, x^1, \ldots, x^r, \ldots,$ in $\{0,1\}^n$ *progresses monotonically towards a limit a* if

$$d(x^{r+1}, a) \lneq d(x^r, a).$$

That is to say, if x^{r+1} has *more* components equal to the corresponding components of a than x^r.

(Clearly, if one progresses monotonically towards a root, then the root is reached in at most n steps!)

Theorem 3. *Let a be a root of F and let x^r be the r-th iterate in the sequence*

$$x^{r+1} = x^r \oplus [A_r^{-1} \otimes F(x^r)].$$

Furthermore, let $[x^r, a] = [x^r, u_1, \ldots, u_{s-1}, a]$ be a chain of minimal length linking x^r to a (see preceding chapter).

x^r and u_1 are neighbours, so are u_1 and u_2, u_2 and u_3 and so on until u_{s-1} and a. The chain so chosen is of minimal length, but not unique. There are therefore indices $j_1, j_2, \ldots, j_s$ mutually different such that

$$\begin{aligned} u_1 &= x^r \oplus e_{j_1} \\ u_2 &= u_1 \oplus e_{j_2} \\ &\vdots \\ u_{s-1} &= u_{s-2} \oplus e_{j_{s-1}} \\ a &= u_{s-1} \oplus e_{j_s}. \end{aligned}$$

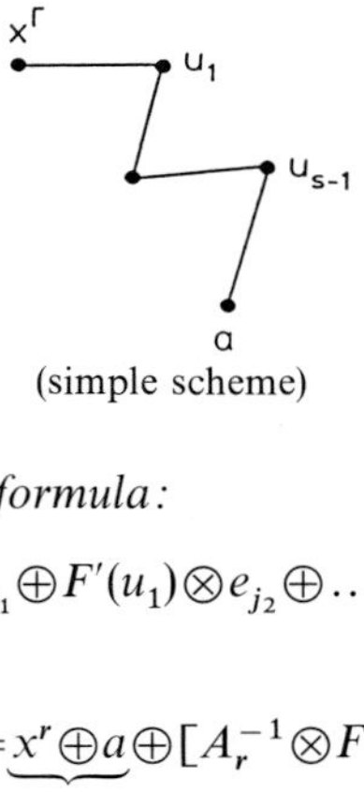

(simple scheme)

Then one has the following correction formula:

$$d(x^{r+1}, a) = d(x^r, a) \oplus [A_r^{-1} \otimes [F'(x^r) \otimes e_{j_1} \oplus F'(u_1) \otimes e_{j_2} \oplus \ldots \oplus F'(u_{s-1}) \otimes e_{j_s}]]. \quad \text{(F)}$$

Indeed

$$d(x^{r+1}, a) = x^{r+1} \oplus a = \underbrace{x^r \oplus a}_{d(x^r, a)} \oplus [A_r^{-1} \otimes F(x^r)].$$

Now

$$\begin{aligned} F(x^r) &= F(x^r) \oplus F(a) \\ &= \underbrace{F(x^r) \oplus F(u_1)}_{} \oplus \underbrace{F(u_1) \oplus F(u_2)}_{} \oplus \ldots \oplus \underbrace{F(u_{s-1}) \oplus F(a)}_{} \\ &F'(x^r) \otimes [\underbrace{x^r \oplus u_1}_{e_{j_1}}] \oplus F'(u_1) \otimes [\underbrace{u_1 \oplus u_2}_{e_{j_2}}] \oplus \ldots \oplus F'(u_{s-1}) \otimes [\underbrace{u_{s-1} \oplus a}_{e_{j_s}}] \end{aligned}$$

which proves the result. □

The preceding formula allows us to understand what happens when we go from x^r to x^{r+1}. $d(x^r, a)$ is a boolean vector having its components equal to 1 (precisely indexed by $j_1, j_2, \ldots, j_s$) corresponding to components of x^r different from those of a. One may schematize this thus (letting $J = \{j_1, \ldots, j_s\}$)

$$d(x^r, a) = \begin{bmatrix} 0 \\ \hline 1 \\ \vdots \\ 1 \\ \hline 0 \end{bmatrix} \left.\begin{matrix} \\ \\ \\ \end{matrix}\right\} J.$$

Then $d(x^{r+1}, a)$ is obtained by adding ($\oplus$) to this vector the following vector:

$$A_r^{-1} \otimes [F'(x^r) \otimes e_{j_1} \oplus F'(u_1) \otimes e_{j_2} \oplus \ldots \oplus F'(u_{s-1}) \otimes e_{j_s}].$$

One may then deduce a necessary and sufficient condition for the monotonic progression of the algorithm.

Corollary. *A necessary and sufficient condition for*

$$d(x^{r+1}, a) \lneq d(x^r, a) \quad \textit{is:}$$

i) $w = F'(x^r) \otimes e_{j_1} \oplus F'(u_1) \otimes e_{j_2} \oplus \ldots \oplus F'(u_{s-1}) \otimes e_{j_s} \neq 0.$

ii) A_r *is such that* $A_r^{-1} \otimes w$ *remains in the space generated by* $e_{j_1}, e_{j_2}, \ldots, e_{j_s}$.

Indeed, to say that $d(x^{r+1}, a) \lneq d(x^r, a)$ is the same as to say that $d(x^{r+1}, a)$ already has the same zero components as $d(x^r, a)$ as well as *at least one additional zero component* (which is necessarily indexed by J). In order that this should be the case, it is necessary and sufficient (when passing from $d(x^r, a)$ to $d(x^{r+1}, a)$) that we add a *non-zero* vector to $d(x^r, a)$ whose non-zero components are indexed from J. This is then the same as conditions i) and ii). □

These conditions are difficult to use in practice. Nevertheless, we will explicitly state some very particular cases where they are verified.

Theorem 4. *If, at stage r, we have*

$$\begin{cases} A_r \otimes e_{j_1} = F'(x^r) \otimes e_{j_1} \\ A_r \otimes e_{j_2} = F'(u_1) \otimes e_{j_2} \\ \vdots \\ A_r \otimes e_{j_s} = F'(u_{s-1}) \otimes e_{j_s} \end{cases}$$

(that is to say that if A_r *has the same* j_1*-th column as* $F'(x_r)$*, the same* j_2*-th column as* $F'(u_1)$ *etc. ..., the same* j_s*-th column as* $F'(u_{s-1})$*) and if* A_r *is non-*

singular, then

$$x^{r+1}=a$$

is the desired root.

Indeed, the vector $A_r^{-1}\otimes w$ then coincides with $e_{j_1}\oplus e_{j_2}\oplus\ldots\oplus e_{j_s}=d(x^r, a)$ from which, according to the formula (F) $d(x^{r+1}, a)=0$ and $x^{r+1}=a$. □

Remark. The above conditions then tell us how to choose A_r in order to arrive at the root a in one step. They are unfortunately not usable in practice since one knows neither a nor the set $J=\{j_1, j_2, \ldots, j_s\}$.

Moreover, along the same lines, it is sufficient that A_r satisfies $A_r\otimes[x^r\oplus a]=F(x^r)$ in order that $x^{r+1}=a$.!!!

Here, however, is an example illustrating the preceding theorem.

Example. $n=3$, F is given by the table

	x			$F(x)$		
u_2	0	0	0	0	1	1
u_1	0	0	1	0	1	0
	0	1	0	0	1	1
x^r	0	1	1	1	1	0
a	1	0	0	0	0	0
	1	0	1	1	0	1
	1	1	0	0	1	0
	1	1	1	1	0	0

There is one root $a=\begin{bmatrix}1\\0\\0\end{bmatrix}$.

Let us take for example $x^r=\begin{bmatrix}0\\1\\1\end{bmatrix}$ and let us choose the chain $[x^r, u_1, u_2, a]$ linking x^r to a with $u_1=\begin{bmatrix}0\\0\\1\end{bmatrix}$ and $u_2=\begin{bmatrix}0\\0\\0\end{bmatrix}$.

We obtain $J=\{2, 3, 1\}$ (with $\uparrow j_1$, $\uparrow j_2$, $\uparrow j_3$ under 2, 3, 1 respectively)

$$x^r=\begin{bmatrix}0\\1\\1\end{bmatrix}, \quad F(x^r)=\begin{bmatrix}1\\1\\0\end{bmatrix}, \quad F'(x^r)=\begin{bmatrix}0&1&1\\1&0&0\\0&0&1\end{bmatrix},$$

$$u_1 = \begin{bmatrix} 0 \\ 0 \\ 1 \end{bmatrix}, \qquad F'(u_1) = \begin{bmatrix} 1 & 1 & 0 \\ 1 & 0 & 0 \\ 1 & 0 & 1 \end{bmatrix},$$

$$u_2 = \begin{bmatrix} 0 \\ 0 \\ 0 \end{bmatrix}, \qquad F'(u_2) = \begin{bmatrix} 0 & 0 & 0 \\ 1 & 0 & 0 \\ 1 & 0 & 1 \end{bmatrix}.$$

A_r now must have the same 2-nd column as $F'(x^r)$, the same 3-rd column as $F'(u_1)$, the same 1-st column as $F'(u_2)$ from which we obtain

$$A_r = \begin{bmatrix} 0 & 1 & 0 \\ 1 & 0 & 0 \\ 1 & 0 & 1 \end{bmatrix} \quad A_r \text{ is regular.}$$

We solve $A_r \otimes z = F(x^r)$ from which $z = \begin{bmatrix} 1 \\ 1 \\ 1 \end{bmatrix}$ and we obtain

$$x^{r+1} = x^r \oplus z = \begin{bmatrix} 0 \\ 1 \\ 1 \end{bmatrix} + \begin{bmatrix} 1 \\ 1 \\ 1 \end{bmatrix} = \begin{bmatrix} 1 \\ 0 \\ 0 \end{bmatrix}.$$

This is clearly the solution a.

If we now take

$$x^r = \begin{bmatrix} 1 \\ 1 \\ 1 \end{bmatrix} \quad \text{with} \quad u_1 = \begin{bmatrix} 1 \\ 0 \\ 1 \end{bmatrix} \quad (\text{chain } [x^r, u_1, a])$$

we have $J = \{2, 3\}$

$$F(x^r) = \begin{bmatrix} 1 \\ 0 \\ 0 \end{bmatrix}, \quad F'(x^r) = \begin{bmatrix} 0 & 0 & 1 \\ 1 & 0 & 1 \\ 0 & 1 & 0 \end{bmatrix}, \quad F'(u_1) = \begin{bmatrix} 1 & 0 & 1 \\ 1 & 0 & 0 \\ 1 & 1 & 1 \end{bmatrix}.$$

A_r now has to have the same second column as $F'(x^r)$ and the same third column as $F'(u_1)$. The first column of A_r may be arbitrary (except for the condition that A_r has to be regular).

Solving $A_r \otimes z = F(x^r)$ we obtain, in all the cases that

$$z = \begin{bmatrix} 0 \\ 1 \\ 1 \end{bmatrix} \quad \text{from which} \quad x^{r+1} = x^r \oplus z = \begin{bmatrix} 1 \\ 0 \\ 0 \end{bmatrix}.$$

Particular case 1. *In the standard Newton method* $(A_r = F'(x^r))$

$$\begin{cases} \text{if } x^r \text{ is an immediate neighbour of } a \\ \text{if } F'(x^r) \text{ is regular} \end{cases}$$

then $x^{r+1} = a$.

This means that if one arrives in the immediate neighbourhood of a root a while iterating with the standard Newton method then this root is found in the next step.

Indeed, with the preceding notations, we have

$$u_1 = a \qquad a = x^r \oplus e_{j_1} \qquad \text{as well as}$$

$$d(x^{r+1}, a) = \underbrace{d(x^r, a)}_{e_{j_1}} \oplus \{[F'(x^r)]^{-1} \otimes F'(x^r) \otimes e_{j_1}\} = 0$$

from which $x^{r+1} = a$. □

We may recall Example 1 at the start of this chapter, p. 135, where this situation occurs. In particular we consider the subgraph

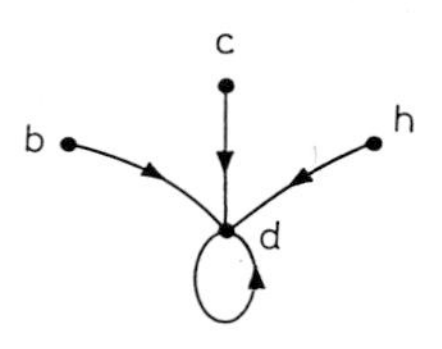

Here d is a root and b, c and h are the only immediate neighbours of the root.

It is also true that the preceding result is valid when $F'(x^r)$ is singular (see p. 164). Here we stay within the confines of an analysis assuming $F'(x^r)$ to be regular.

Corollary. *If two distinct roots* a_1 *and* a_2 *of F have a common immediate neighbour b, then* $F'(b)$ *is not regular.*

Indeed, if the standard Newton method is started at $b = x_0$ and if it is defined ($F'(b)$ invertible) we get both $x_1 = a_1$ and $x_1 = a_2$, which is impossible. The conclusion is therefore clear since $F'(b)$ then has two identical columns (one obtains in effect the same change in F whenever we pass from b to either of its neighbours a_1 and a_2). □

Particular case 2. *If the derivative* F' *is constant (and invertible) in a massive neighbourhood of a root of F, and if* x^r *falls into this massive neighbourhood*

then we have, using the standard Newton method,

$$x^{r+1}=a.$$

Indeed, if x^r is in a massive neighbourhood W_a of a then by definition the chain $[x^r, a]$ is completely contained in W_a. This now results in

$$A_r=F'(x^r)=F'(u_1)=\ldots=F'(u_{s-1}),$$

using the notation of Theorem 3. This verifies the conditions of Theorem 4, from which the result follows. □

As a trivial example of this, one may quote the case where F is affine ($F(x)=[T\otimes x]\oplus u$ with an invertible T). One takes all of $(Z/2)^n$ for W_a. $F'(x)$ is constant and equal to T for all x. Then, independent of which x^0 is chosen, we have $x^1=T^{-1}\otimes u=a$, the unique root of F.

Here is a less trivial example where

$n=3$ and where F is given by the table

	x	$F(x)$
	0 0 0	1 1 1
b	0 0 1	1 1 0
	0 1 0	α β γ
	0 1 1	1 0 1
d	1 0 0	0 0 1
a	1 0 1	0 0 0
	1 1 0	0 1 0
c	1 1 1	0 1 1

α, β, γ may be chosen arbitrarily from $\{0, 1\}$

There is one root $a=(1\ 0\ 1)$. One takes the immediate neighbourhood of a for a massive neighbourhood of a thus obtaining

$$V_a=\{a, b, c, d\}.$$

Clearly,

$$F'(a)=F'(b)=F'(c)=F'(d)=\begin{bmatrix}1 & 0 & 0\\ 1 & 1 & 0\\ 0 & 1 & 1\end{bmatrix} \quad \text{(regular matrix).}$$

One verifies that, starting from a, b, c or d, the standard Newton method leads to the root a in one step.

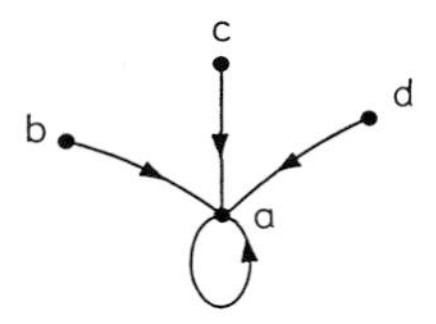

(This example moreover illustrates the particular case 1.)

Remark. In this example, F is not necessarily affine (precisely, if $(\alpha\ \beta\ \gamma)$ is not equal to $(1\ 0\ 0)$ then F is not affine).

It would be interesting to be able to give less drastic conditions for the derivative F' and for the choice of A_r that still guarantee the necessary and sufficient conditions i) and ii) for the monotone behaviour of the method (Theorem 3 and its corollary). The problem of providing such conditions is *unsolved* (and therefore left as an exercise for the reader!).

In the following example, however, the algorithm progresses monotonically.

Example. $n=3$. We take a priori the incidence matrix $B(F)$ to be upper triangular

$$B(F)=\begin{bmatrix}1 & 1 & 1\\ 0 & 1 & 1\\ 0 & 0 & 1\end{bmatrix}$$

with for example (boolean notation)

$$f_1(x)=x_1x_2+x_2x_3$$
$$f_2(x)=\bar{x}_2+x_3$$
$$f_3(x)=x_3.$$

From this we have the table for F

	x	$F(x)$
	0 0 0	0 1 0
	0 0 1	0 1 1
a	0 1 0	0 0 0
	0 1 1	1 1 1
	1 0 0	0 1 0
	1 0 1	0 1 1
	1 1 0	1 0 0
	1 1 1	1 1 1

There is one root $a=(0\ 1\ 0)$.

Let us take the (ultra) simplified Newton method with $A_r=I$ (unit matrix)

$$x^{r+1}=x^r\oplus F(x^r).$$

Then for

$$x^0=\begin{bmatrix}1\\0\\0\end{bmatrix},\qquad d(x^0,a)=\begin{bmatrix}1\\1\\0\end{bmatrix}$$

we get

$$x^1 = x^0 \oplus F(x^0) = \begin{bmatrix}1\\1\\0\end{bmatrix}, \qquad d(x^1, a) = \begin{bmatrix}1\\0\\0\end{bmatrix},$$

$$x^2 = x^1 \oplus F(x^1) = \begin{bmatrix}0\\1\\0\end{bmatrix}, \qquad d(x^2, a) = \begin{bmatrix}0\\0\\0\end{bmatrix}.$$

The convergence is monotonic.

6. The Efficiency of an Iterative Method on a Finite Set

We will now try to be a bit more precise about what may be an "efficient" iteration on a finite set before we examine some more extensive numerical experiments. This point was already touched upon in Chap. 4. (If F is contracting then the associated Gauss-Seidel operator is more contracting or, more generally, if a serial-parallel operator is contracting then all the operators that are more sequential are also more contracting.) In Chap. 5 we were furthermore able to compare parallel processes and serial processes. Given certain hypotheses we could prove that the latter is "better" than the former.

A priori our aim here is to seek a root of F among the 2^n elements of $\{0, 1\}^n$. For this we consider an iteration that is a method of successive approximations for an operator H associated with F

$H(x) = x \oplus [A^{-1} \otimes F(x)]$ for the simplified Newton method,

$H(x) = x \oplus [[F'(x)]^{-1} \otimes F(x)]$ for the standard Newton method,

where the roots of F are characterized as fixed points for H.

Such an iteration is characterized by its graph since it operates on a finite set. This graph is decomposed into connected components called basins (see Chap. 1). A distinction is made between the components that have a fixed point and those that have a cycle (of length ≥ 2).

In our context, the algorithm has a successful completion if, for a given starting point, it ends at a fixed point of the iteration (that is, a root of F). If the algorithm ends at a cycle of length ≥ 2 then this corresponds to a failure of the iteration. Empirically, therefore, the iterative method is considered "good" if its iteration graph has the following properties:

1) Few of the basins have a cycle. Furthermore, if a basin has a cycle then one hopes that this basin contains few points (low frequency of the cycling phenomenon in the iteration).

2) The other basins (which all contain a fixed point of H, that is, a solution to our problem) should be as "close" as possible to their fixed points (fast convergence to the fixed point).

Having stated these notions in a deliberately informal context (since we are unable to define the notion of efficiency in a more rigorous manner), we now examine some numerical results.

7. Numerical Experiments (Executed by M. Jiang Zegu cf. [103])

For $n=5$, the following table was created randomly:

	x					$F(x)$				
0	0	0	0	0	0	1	1	0	0	0
1	0	0	0	0	1	0	0	1	0	0
2	0	0	0	1	0	1	1	1	1	1
3	0	0	0	1	1	1	1	0	0	0
4	0	0	1	0	0	0	1	1	0	1
5	0	0	1	0	1	1	0	0	1	1
6	0	0	1	1	0	0	1	1	1	1
7	0	0	1	1	1	1	1	0	0	1
8	0	1	0	0	0	0	0	0	0	1
9	0	1	0	0	1	0	0	1	1	0
10	0	1	0	1	0	1	1	0	0	1
11	0	1	0	1	1	0	0	1	1	0
12	0	1	1	0	0	0	0	1	0	0
13	0	1	1	0	1	1	0	0	1	0
14	0	1	1	1	0	0	0	1	1	1
15	0	1	1	1	1	0	0	0	1	0
16	1	0	0	0	0	0	0	0	0	1
17	1	0	0	0	1	1	0	1	0	1
18	1	0	0	1	0	1	0	0	0	1
19	1	0	0	1	1	1	0	1	1	1
20	1	0	1	0	0	0	0	0	1	0
21	1	0	1	0	1	0	0	1	1	0
22	1	0	1	1	0	1	0	1	0	0
23	1	0	1	1	1	1	1	0	1	1
24	1	1	0	0	0	1	1	0	0	0
25	1	1	0	0	1	0	1	0	0	0
26	1	1	0	1	0	1	1	1	1	0
27	1	1	0	1	1	1	0	1	0	1
28	1	1	1	0	0	1	1	1	1	0
29	1	1	1	0	1	1	1	0	0	1
30	1	1	1	1	0	1	0	1	0	0
31	1	1	1	1	1	1	1	1	0	0

This function, which has no roots, is called F_0.

F_1 is obtained by modifying the table entry $x=(0\ 0\ 0\ 1\ 0)$ such that $F_1(0\ 0\ 0\ 1\ 0)=(0\ 0\ 0\ 0\ 0)$. This means that F_1 has one root.

F_2 is obtained by modifying F_1 only for $x=(1\ 0\ 0\ 0\ 0)$ such that $F_2(1\ 0\ 0\ 0\ 0)=(0\ 0\ 0\ 0\ 0)$. This means that F_2 has two roots.

F_3 is obtained by modifying the table for F_2 for $x=(1\ 1\ 0\ 1\ 1)$ and obtaining $F_3(1\ 1\ 0\ 1\ 1)=(0\ 0\ 0\ 0\ 0)$. Then F_3 has three roots.

In the same manner F_5, F_8 and F_{10} are defined.

In the following pages the iteration graphs for the standard Newton and simplified Newton methods are given for F_0, F_1, F_2, F_3, F_5, F_8 and F_{10}.

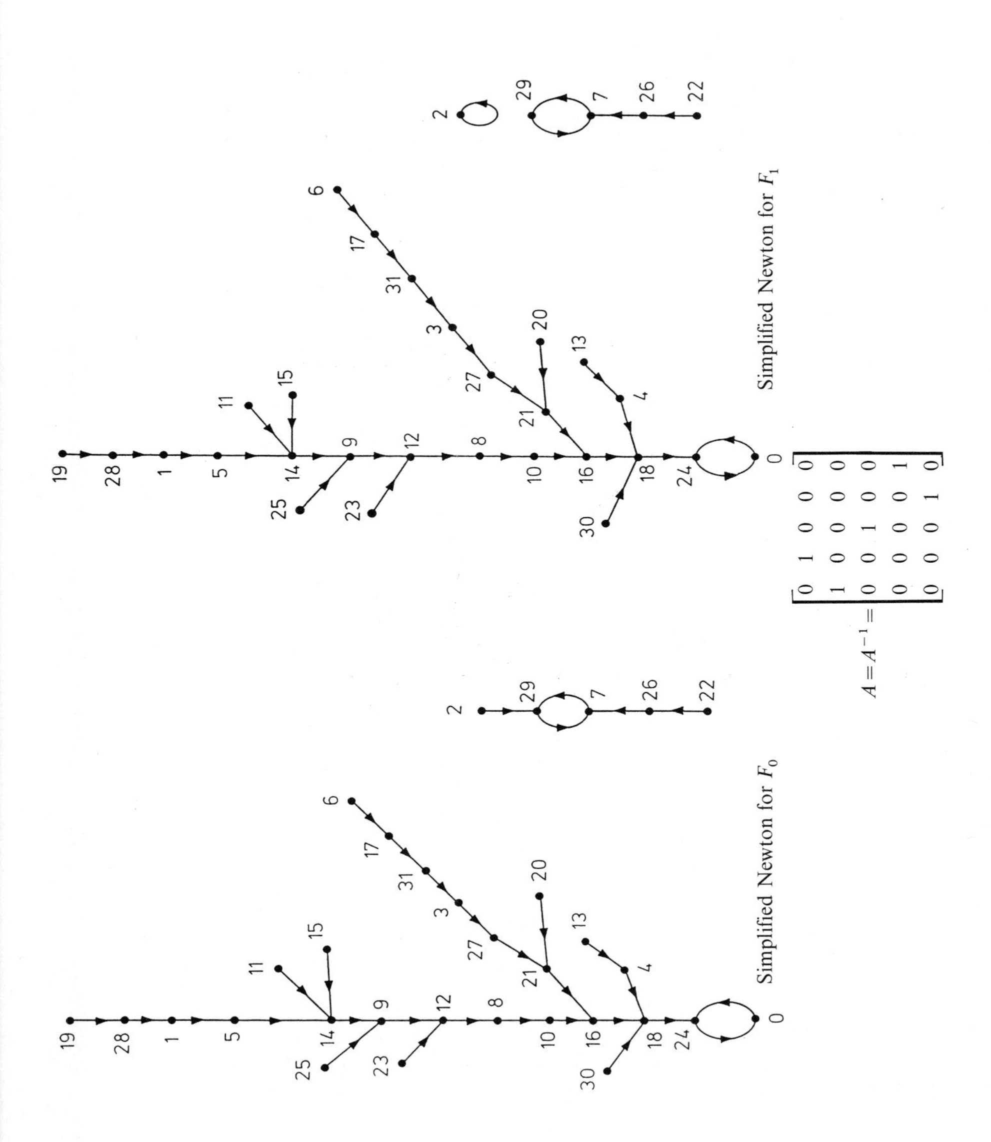

Simplified Newton for F_0

Simplified Newton for F_1

$$A=A^{-1}=\begin{bmatrix} 0 & 1 & 0 & 0 & 0 \\ 1 & 0 & 0 & 0 & 0 \\ 0 & 0 & 1 & 0 & 0 \\ 0 & 0 & 0 & 0 & 1 \\ 0 & 0 & 0 & 1 & 0 \end{bmatrix}$$

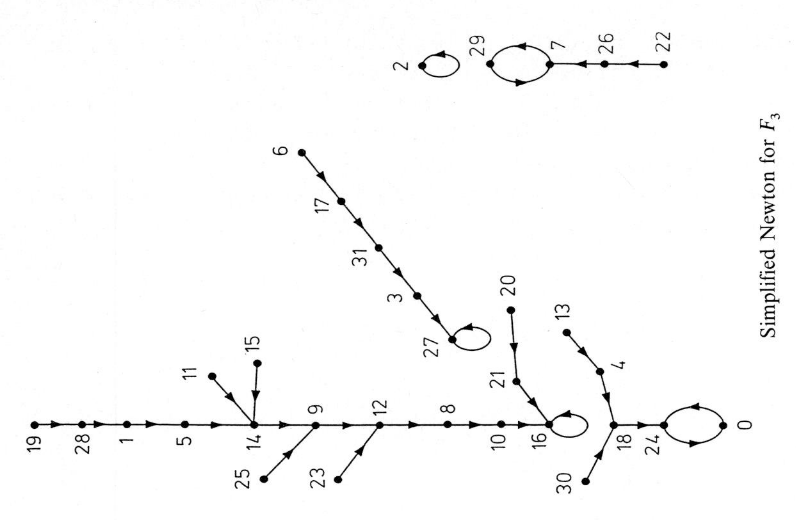

Simplified Newton for F_3

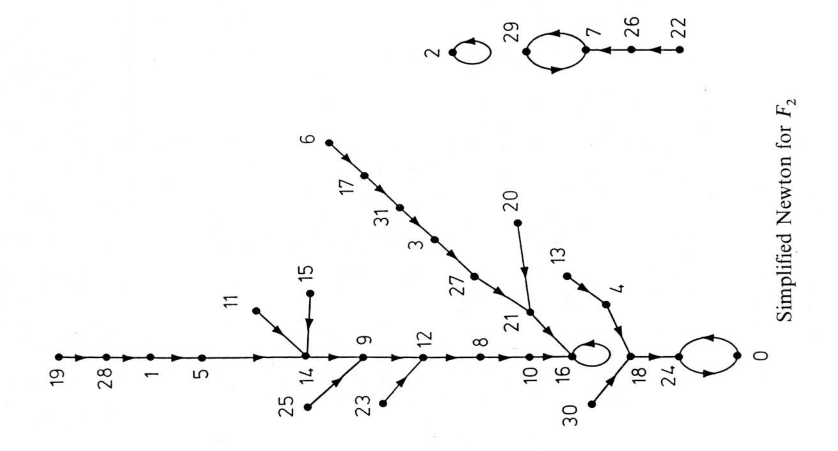

Simplified Newton for F_2

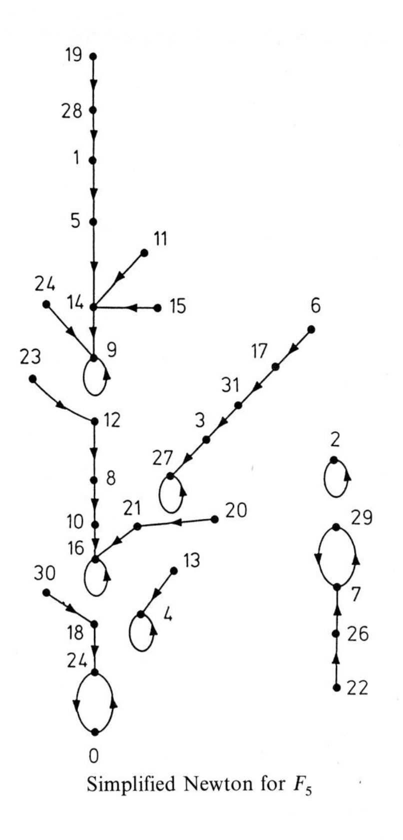

Simplified Newton for F_5

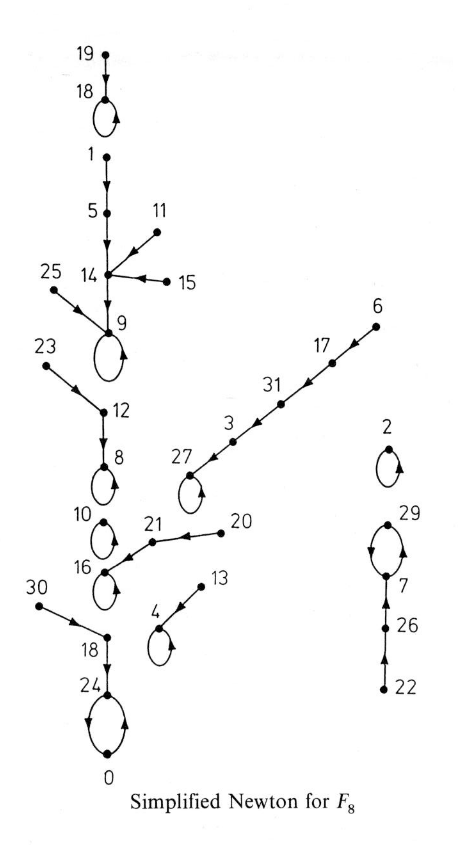

Simplified Newton for F_8

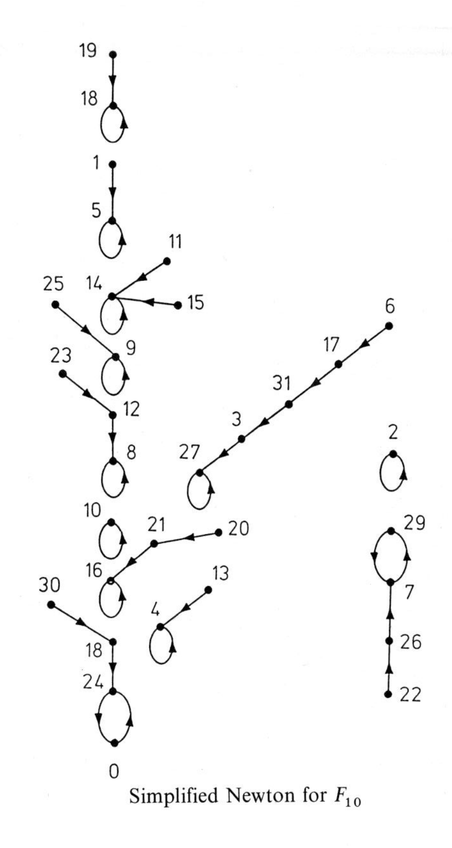

Simplified Newton for F_{10}

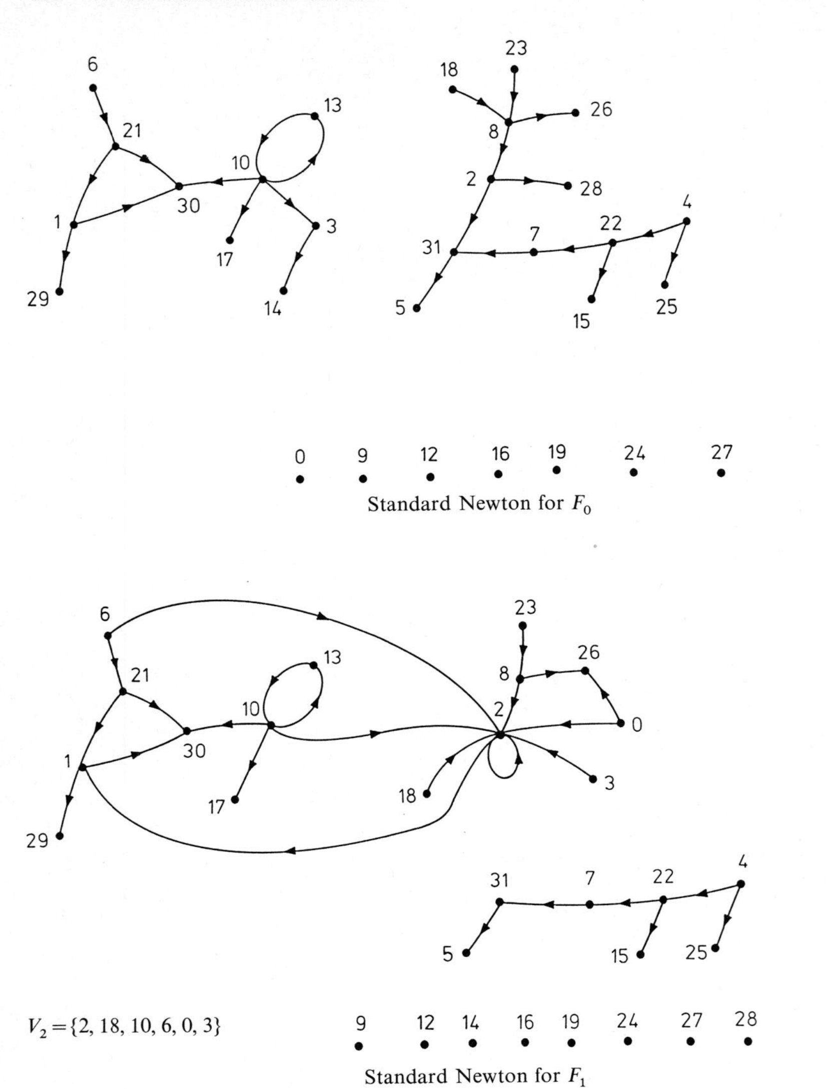

Standard Newton for F_0

$V_2 = \{2, 18, 10, 6, 0, 3\}$

Standard Newton for F_1

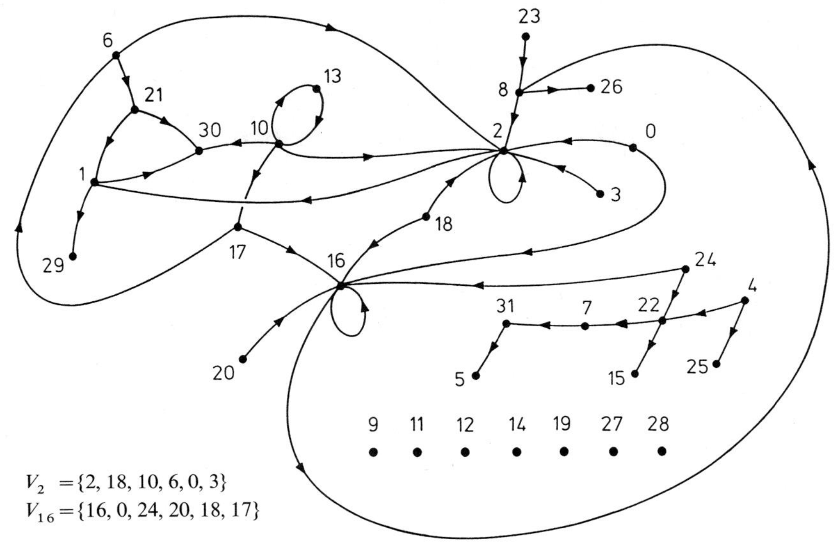

Standard Newton for F_2

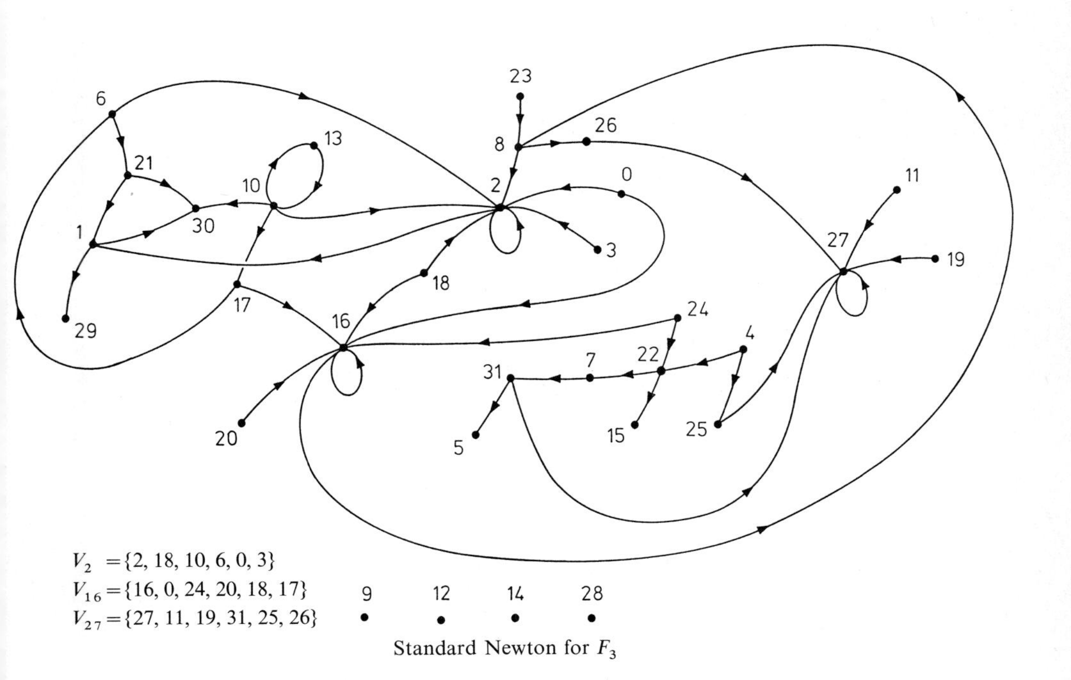

Standard Newton for F_3

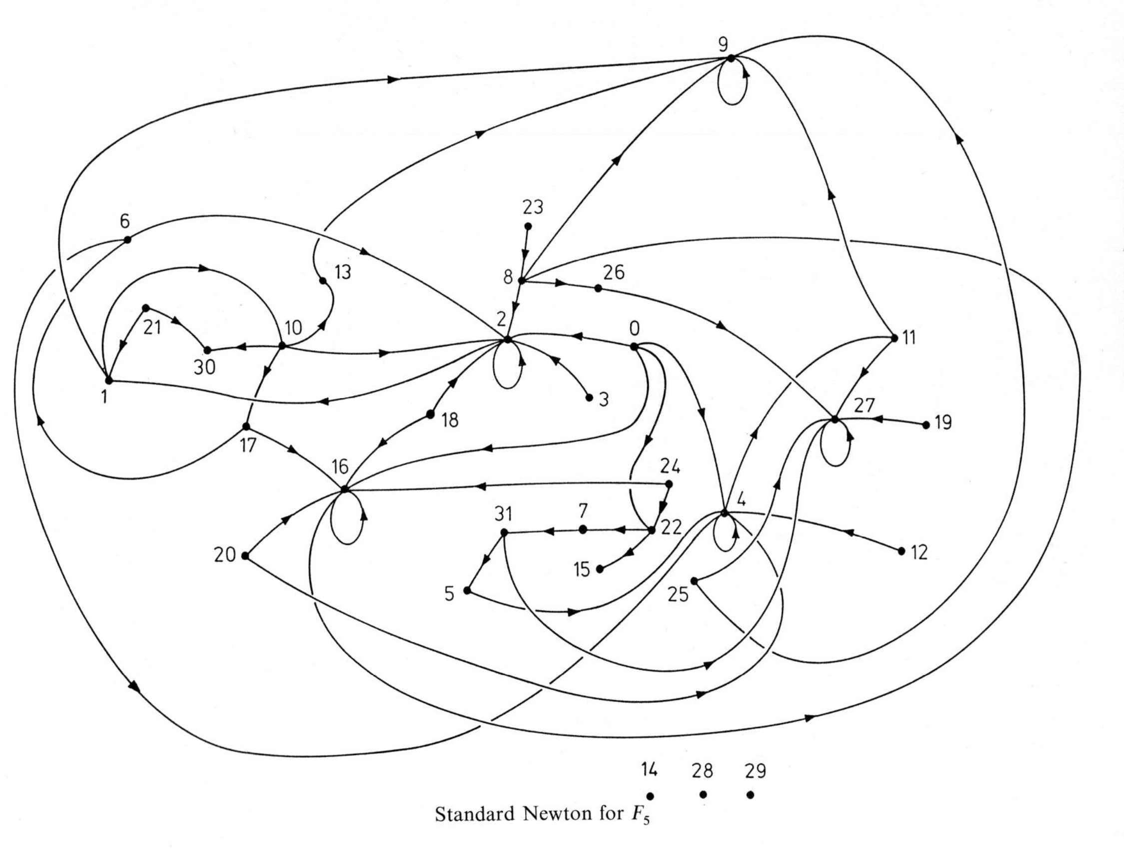

Standard Newton for F_5

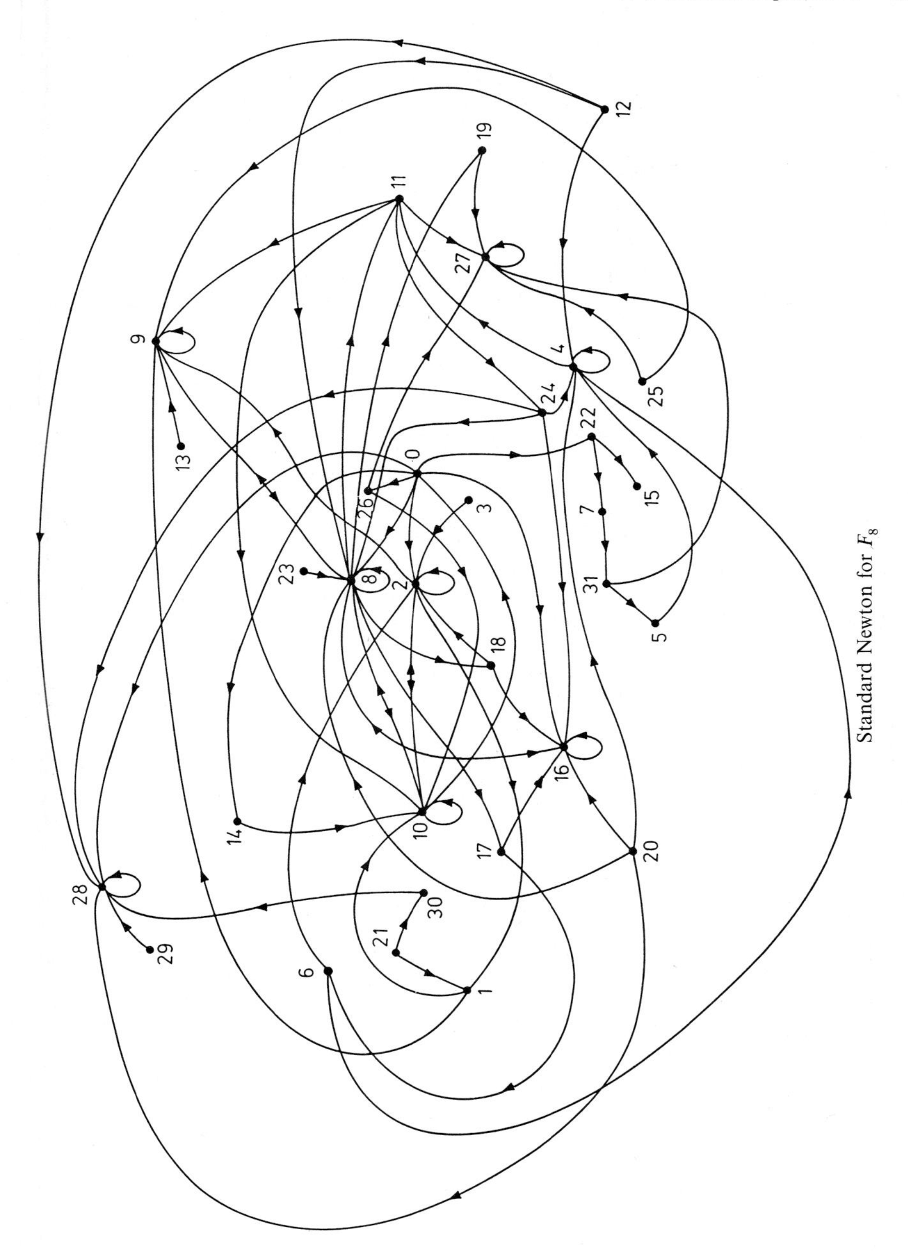

Standard Newton for F_8

RECAPITULATION

		The number of points where the iteration leads to			Number of cycles	Maximum length of the cycles	Number of points leading to a cycle	Number of wells (impossible systems)	Maximum length of an iteration leading to		
		A root	Possibly a root	Not a root					A fixed point	A cycle	A well
Simplified Newton	F_0	0	0	32	2	2	32	0	/	11	/
	F_1	1	0	31	2	2	31	0	0	11	/
	F_2	22	0	10	2	2	10	0	9	3	/
	F_3	22	0	10	2	2	10	0	9	3	/
	F_5	24	0	8	2	2	8	0	5	2	/
	F_8	24	0	8	2	2	8	0	4	2	/
	F_{10}	24	0	8	2	2	8	0	4	2	/
Standard Newton	F_0	0	0	32	1	2	2	17	/	0	4
	F_1	3	6	23	1	2	2	16	2	0	4
	F_2	5	8	19	1	2	2	13	4	0	4
	F_3	13	11	8	1	2	2	7	4	0	4
	F_5	21	6	5	0	/	/	5	5	/	2
	F_8	29	2	1	0	/	/	1	6	/	1
	F_{10}	32	0	0	0	/	/	0	6	/	/

Standard Newton: observed rate of regular matrices			Theoretical rate $\tau_5^2=0.298$ see Appendix 2
	F_0	0.250	
	F_1	0.218	
	F_2	0.187	
	F_3	0.281	
	F_5	0.281	
	F_8	0.312	

Comments. Upon re-examination of the various iteration graphs and tables that we have displayed we make the following observations:

– First of all it is interesting to follow the development of these graphs as a function of the increasing number of roots of the functions F_i considered ($F_0, F_1, F_2, F_3, F_5, F_8, F_{10}$).

Indeed, one observes that certain parts of the graphs are conserved, while other parts are modified. This may be made more precise.

This evolution of the iteration graphs is evident in the simplified Newton method, already starting from the graph related to F_0. Each new root that is introduced "breaks" the iteration graph at the root, while the rest of the graph is unchanged. This is also a fact visible in the expression for the iteration since for all x^r that is not a newly introduced root the iteration

$$x^{r+1} = x^r \oplus [A^{-1} \otimes F(x^r)]$$

is not modified. There is indeed always an arrow from x^r to x^{r+1} in the iteration graph.

This development is more complex for the standard Newton method. Whenever a new root is introduced into the function, the only arrows that are modified (this was to be expected from the expression for the iteration*) are the arrows issuing from the injected root and from its first neighbours. This development will be made more precise at the end of this chapter.

– Having said this, we note that the two methods work rather well. There is relatively often convergence to a root (above all in the standard Newton method), and the phenomenon of cycling does not occur often. As one might expect, moreover, the two methods work better the more roots F has. For the standard Newton method we also have that if it converges, then the convergence is fast (this is due to the local information carried by $F'(x^r)$).

– There is no problem in passing from x^r to x^{r+1} in the simplified Newton method. We simply choose A_r regular such that x^{r+1} is always defined. *This is not the case with the standard Newton method. We note, in fact, that there is a very high occurrence of singular matrices (of the order of 70% to 80%). This fact complicates both the calculations, the looks of the graphs and their analysis. Whenever the system $F'(x^r) \otimes z = F(x^r)$ is singular one still tries to solve it*, however, getting either many solutions (2^k) or an impossibility.

In fact, in the iteration graphs for the standard Newton method there are points that are wells *(*impossible systems*)* and numerous points that have many arrows (2^k) issuing from them (many possible solutions of the singular system).

---

* Indeed, at a point x^r that is not a neighbour to the injected root we have that $F'(x^r)$ is unchanged. This means that there is also no change in the system $F'(x^r) \otimes z = F(x^r)$ and hence the network of arrow(s) leaving x^r is unchanged.

– This discovery, that the standard Newton method is rather strongly penalized in $(Z/2)^n$ leads us to ask the following question: *What percentage* τ_n^p *of* $n \times n$ *matrices with elements in* Z/p *turn out to be regular* (p *prime*)?

A simple calculation (see Appendix 2) shows that if p is small then this rate is indeed rather small for all n. For example $\tau_5^2 = 0.298$ (less than 30%). In fact, in our examples (where $p=2$ and $n=5$) the frequency of regular matrices is observed to be of this order as shown by the following numerical results:

F_0	F_1	F_2	F_3	F_5	F_8	τ_5^2
0.250	0.218	0.187	0.281	0.281	0.312	0.298

For a fixed n we have that τ_n^p increases rapidly towards 1 as p increases (see Appendix 2). It might therefore be interesting (see conclusions) to redo this study in $(Z/p)^n$ and to conduct further experiments in this space. For $p = 7$, one might have the order of 20% singular matrices as opposed to the 80% rate of singular matrices observed here. This study has indeed been executed in [79], and the results confirm the above estimate.

– Having stated this, let us now place ourselves at a root a of an arbitrary F. *On the graph of the standard Newton method for this* F *we observe that all the immediate neighbours of* a, *denoted by* x^r, *have an arrow leading to* a. This is moreover shown as follows by the fact that we have

$$F(x^r) = F(x^r) \oplus F(a) = F'(x^r) \otimes [x^r \oplus a].$$

Thus $z = x^r \oplus a$ is a solution of the system $F'(x^r) \otimes z = F(x^r)$ from which we have the result that an arrow leaves x^r to go directly to $x^r \oplus z = a$.

If $F'(x^r)$ *is regular then this arrow is the unique arrow leaving* x^r (see § 5 above, particular case 1). *If* $F'(x^r)$ *is singular, then there exist other arrows leaving* x^r (*for other points*).

Similarly, if $F'(a)$ *is regular, then a unique arrow leaves* a *and loops back to* a. *If* $F'(a)$ *is singular then several arrows leave* a, *however, one loops back to* a. *These points are all verified in the examples considered above.*

– As a concluding remark for this numerical experimentation *we may state that the standard Newton method in* $(Z/2)^n$ *allows us to reach a root of* F *rather often* (*few cycles*) *and fast*. This is also confirmed by the similar behaviour of many other examples (see Appendix 3). *The standard Newton method is, however, encumbered by the very high rate* (*the order of* 80%) *of singular matrices* $F'(x^r)$. *This both causes difficulties in the numerical implementation as well as sometimes leads to impossible systems.*

These inconveniences are not present in the simplified Newton method. This method is, however, made less interesting by the large number of iteration steps required for convergence. Moreover, we also note the difference in appearance of the graphs for the standard Newton and the simplified Newton methods.

8. Conclusions

We have been able to transpose Newton's method for finding a root of the equation $F(x)=0$ from the framework of iterations in $\mathbb{R}^n$ into the context of discrete iterations in $(Z/2)^n$. Having made this transposition (which was made possible through the use of the notion of a discrete derivative) we were able to study two aspects of the Newton method in $(Z/2)^n$:

(A) Theoretical Analysis of Convergence

(simplified Newton or the general case)

Even though we were not able to transpose the analysis of Kantorovitch into the discrete context, we were able to study the convergence of the method (attractive roots, monotone convergence).

This analysis leads to several remarks.

a) The study within the discrete framework is relatively simpler than the study in $\mathbb{R}^n$ (or in a Banach space). Here the partial convergence results that are obtained correspond to *stationary* convergence (root obtained in no more than n steps). *There are, moreover, no stability problems when executing the numerical experiments since they are executed using exact arithmetic.* Clearly, these results (which might have been developed further) remain rather theoretical, analogous to the results obtained in the classical Newton analysis in $\mathbb{R}^n$. The fact is, that the conditions that are manipulated require the knowledge of the root that is being searched for, or the knowledge of some rather inaccessible conditions. In a realistic case this is not possible.

b) It would be interesting to complete the convergence studies by investigating iterations that end up in a cycle. This is, however, difficult!

c) Finally, we studied the case of A_r being regular for simplicity. This is frequently not the case when the standard Newton method is applied in practice.

We therefore point out that the study can be redone without large modifications when A_r is singular (x^{r+1} is then not uniquely defined, and may also not be defined at all). For example, formula (F) of Theorem 3 now reads as follows:

$$A_r \otimes d(x^{r+1}, a) = A_r \otimes d(x^r, a) \oplus [F'(x^r) \otimes e_{j_1} \oplus \dots F'(u_{s-1}) \otimes e_{j_s}].$$

(B) Numerical Experimentation

In the classical Newton method there is a certain gap between the theoretical studies of convergence and the convergence observed when experimenting with the method. Indeed, the method converges more often than one could predict from theory. The same is in fact the case in the discrete setting, where (fortunately!) the numerically convergent cases to a large extent lie outside the bounds of the theoretically convergent cases. This means that the method has a certain practical efficiency that has not been formalized. (See also the Introduction.)

The frequent occurrence of a singular system makes the method difficult to implement. This difficulty is strongly reduced in $(Z/p)^n$ when p is greater than 6. This leads us to the idea of reconsidering the method in $(Z/p)^n$. In a similar manner to what was done for the method in $(Z/2)^n$ one would define, analyze the behaviour (convergence, cycles) and perform experiments. Much of this work has been done and we refer to [79] for an account of it.

General Conclusion

In this monograph the behaviour of discrete iterations has been studied. *By introducing a metric tool it was possible to perform a non-trivial analysis of the behaviour of these iterations obtaining some interesting results.* These results were mainly obtained by transposing results in continuous metric spaces into the discrete context.

This transposition was carried furthest in the present chapter, where a Newton-like method was defined based on results from previous chapters as well as on Newton's method for Banach spaces. Additionally, a numerical analysis was performed on this method within the discrete context.

It is not at all claimed that the metric approach presented here is the only possible approach to the study of discrete iterations. Indeed, for other approaches, see the lengthy bibliography, in particular Sect. D, Refs. [63] to [141], especially mentioning Refs. [139]–[141] and also Ref. [191].

Appendix 1

The Number of Maps of $\{0, 1\}^n$ into $\{0, 1\}^n$

(A) There are clearly $(2^n)^{2^n}$ maps of $\{0, 1\}^n$ into itself

n	1	2	3	4	5
number	4	256	$1.677 \cdot 10^7$	$1.844 \cdot 10^{19}$	$1.461 \cdot 10^{48}$

(B) Among these maps, there are $(2^n - 1)^{2^n}$ which have *no roots*, from which we have the ratio

$$v = \left(1 - \frac{1}{2^n}\right)^{2^n}$$

of maps with no roots to all the maps of $\{0, 1\}^n$ into itself

n	1	2	3	4	5	∞
number	1	81	$5.764 \cdot 10^6$	$6.568 \cdot 10^{18}$	$5.291 \cdot 10^{47}$	
ratio v	0.25	0.316	0.343	0.356	0.362	$1/e = 0.367$

(C) There are therefore $(2^n)^{2^n} - (2^n - 1)^{2^n}$ maps that have *at least one root*

n	1	2	3	4	5	∞
number	3	175	$1.101 \cdot 10^7$	$1.187 \cdot 10^{19}$	$9.323 \cdot 10^{47}$	
ratio δ	0.75	0.683	0.656	0.643	0.637	$0.632 = 1 - 1/e$

Therefore for all n the probability that an arbitrary map of $\{0,1\}^n$ into itself has a root is clearly larger than the probability of having no root.

(D) The number of functions mapping $\{0,1\}^n$ into $\{0,1\}^n$ that have *exactly one root* is

$$2^n(2^n-1)^{2^n-1}$$

and the corresponding ratio τ_1 is

$$\tau_1 = \left(1-\frac{1}{2^n}\right)^{2^n-1} \qquad (\tau_1 \xrightarrow[(n\to\infty)]{} 1/e)$$

n	1	2	3	4	5	∞
number	2	108	$6.588 \cdot 10^6$	$7.006 \cdot 10^{18}$	$5.462 \cdot 10^{47}$	
ratio τ_1	0.5	0.421	0.392	0.379	0.373	$1/e = 0.367$

(E) Indeed, the number of functions of $\{0,1\}^n$ into $\{0,1\}^n$ that have *exactly q roots* is

$$C^q_{2^n}(2^n-1)^{2^n-q}.$$

From this we have the corresponding ratio τ_q as

$$\tau_q = \frac{C^q_{2^n}(2^n-1)^{2^n-q}}{(2^n)^{2^n}} \qquad \left(\tau_q \xrightarrow[(n\to\infty)]{} \frac{1}{q!\,e}\right).$$

(F) In particular, the number of functions having *exactly two roots* is

$$2^{n-1}(2^n-1)^{2^n-1}$$

which is for all n exactly half the number of functions having *one* root

$$\tau_2 = \tfrac{1}{2}\tau_1$$

n	1	2	3	4	5	∞
number	1	54	$3.294 \cdot 10^6$	$3.503 \cdot 10^{18}$	$2.731 \cdot 10^{47}$	
ratio τ_2	0.25	0.210	0.196	0.189	0.186	$1/2e = 0.183$

Finally, we have "for n infinite" that

$$\underset{F\text{ arbitrary}}{\underset{\uparrow}{1}} = \underset{\text{no roots}}{\underset{\uparrow}{\frac{1}{e}}} + \underbrace{\underset{\text{1 root}}{\underset{\uparrow}{\frac{1}{e}}} + \underset{\text{2 roots}}{\underset{\uparrow}{\frac{1}{2e}}} + \underset{\text{3 roots}}{\underset{\uparrow}{\frac{1}{3!\,e}}} + \dots + \underset{q\text{ roots}}{\underset{\uparrow}{\frac{1}{q!\,e}}}}_{\text{at least one root}} + \dots$$

We may therefore display the following graph of the proportion of roots as a function of n:

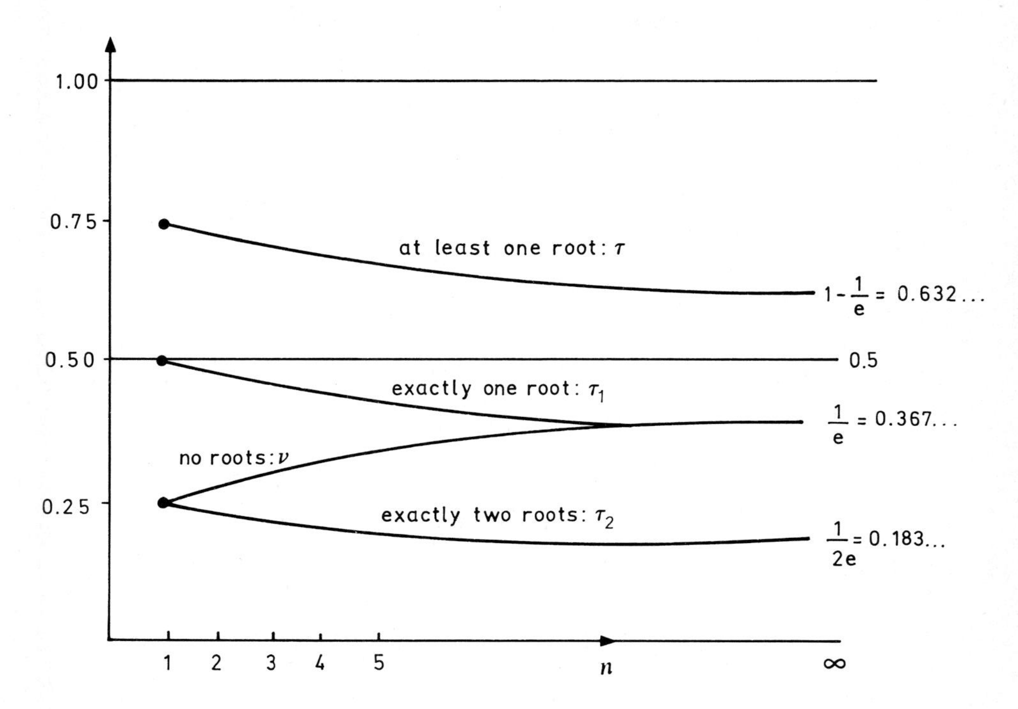

We remark here that one may replace the property "having one root" by "having a fixed point" (or in general "reach a fixed value in $\{0, 1\}^n$") without modifying the preceding results.

Appendix 2

The Number of Regular $n \times n$ Matrices With Elements in Z/p (p prime)

Among the p^{n^2} $n \times n$ matrices having elements from Z/p (p prime) there are exactly

$$(p^n - 1)(p^n - p) \dots (p^n - p^{n-1})$$

that are regular. From this we have that the ratio of regular matrices to all possible $n \times n$ matrices with elements in Z/p is given by

$$\tau_n^p = \left(1 - \frac{1}{p^n}\right)\left(1 - \frac{1}{p^{n-1}}\right) \dots \left(1 - \frac{1}{p}\right).$$

The following table contains evaluations of this ratio for a number of values of n, p.

p \ n	1	2	3	4	5	6	7	8	9	10	∞
2	0.5	0.375	0.328	0.307	0.298	0.293	0.291	0.289	0.289	0.289	0.288
3	0.666	0.592	0.570	0.563	0.561	0.560					0.560
5	0.8	0.768	0.761	0.760							0.760
7	0.857	0.839	0.837	0.836							0.836
11	0.909	0.901	0.900								0.900
13	0.923	0.917									0.917
17	0.941	0.937									0.937
19	0.947	0.944									0.944
97	0.989										0.989
103	0.990										0.990

We see that

– for all p, τ_n^p converges very fast towards a limit for increasing n (in all the cases the limit is reached with a tolerance of 10^{-3} after $n=10$).

– For small p, the ratio of regular matrices to all possible matrices is small. Example: $p=2$, $n=5$, $\tau_5^2=0.298$. This ratio increases rapidly as p increases. We note that for $p=7$ for example, it is greater than 0.8 for all n.

– Finally for large p this ratio rapidly approaches 1.

We now consider the linear systems $Ax=b$ where the $n\times n$ elements of A and the $n\times 1$ elements of b are taken from Z/p (p prime).* The ratio of *solvable* systems to *all possible* systems is denoted by λ_n^p. The following table giving numerical values for λ_n^p for various n and p is based on results obtained in the thesis of M. Jiang Zegu [103].

p \ n	1	2	3	4	5	6	∞
2	0.75	0.672	0.634	0.625	0.617	0.613	0.608
3	0.778	0.726	0.71	0.705	0.703		0.702
5	0.84	0.814	0.809	0.808			0.808
7	0.878	0.863	0.860				0.860
11	0.917	0.911	0.910				0.910
13	0.929	0.924					0.924
17	0.945	0.942					0.942
19	0.950	0.949					0.949
97	0.990						0.990
103	0.990						0.990

The comparison of this table with the preceding one is quite interesting. Indeed, we note that the proportion of linear *solvable* systems (regular or singular) is always significantly larger than the proportion of *regular* systems when p is small. Moreover the behaviour of λ_n^p when n and p increases is rather comparable to that of τ_n^p. For $p>19$, λ_n^p and τ_n^p are quite close for all n.

* For n and p fixed, there are exactly p^{n^2+n} such systems.

Appendix 3

Some Further Examples Illustrating the Standard Newton Method in $(Z/2)^n$ and $(Z/3)^n$

These examples are from the thesis of M. Jiang Zegu [103]. For a study of the method in $(Z/p)^n$ see also S. E. Bernoussi [79].

$$F(x)=\begin{cases} f_1=x_2+1 \\ f_2=x_2x_3+x_3x_4+x_3+1 \\ f_3=f_2 \\ f_4=f_1 \end{cases}$$

$$F\colon (Z/p)^n \to (Z/p)^n, \qquad p=2,\ n=4$$

standard Newton

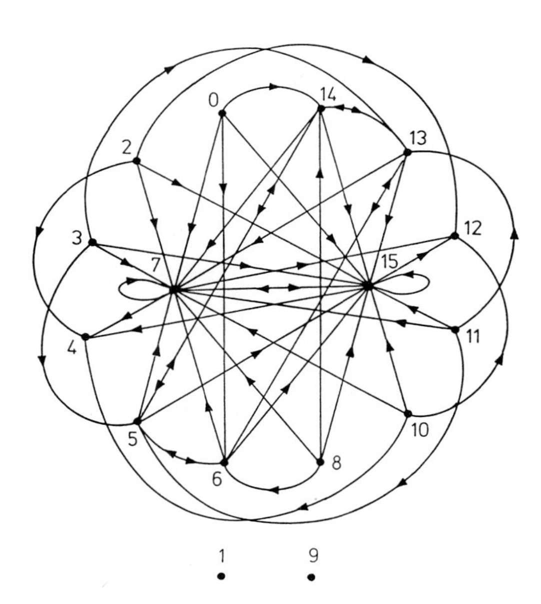

$$F(x)=\begin{cases} f_1=x_2+1 \\ f_2=x_2x_3+x_3x_4+x_3+1 \\ f_3=x_1+x_2x_4+x_3 \\ f_4=f_1 \end{cases}$$

$$F:\ (Z/p)^n \to (Z/p)^n, \qquad p=2,\ n=4$$

standard Newton

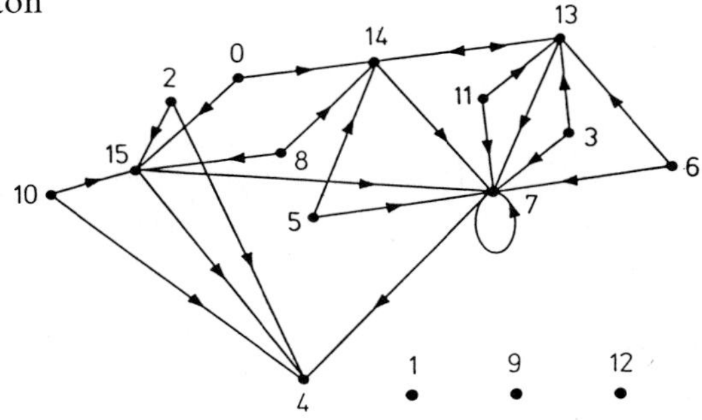

$$F(x)=\begin{cases} f_1=x_2+x_3+x_4+x_5 \\ f_2=x_2+x_3+x_4+x_5+x_1x_2 \\ f_3=x_3+x_4+x_5+x_1x_2x_3 \\ f_4=x_4+x_5+x_1x_2x_3x_4+1 \\ f_5=x_5+x_1x_2x_3x_4x_5 \end{cases} \qquad p=2,\ n=5$$

standard Newton

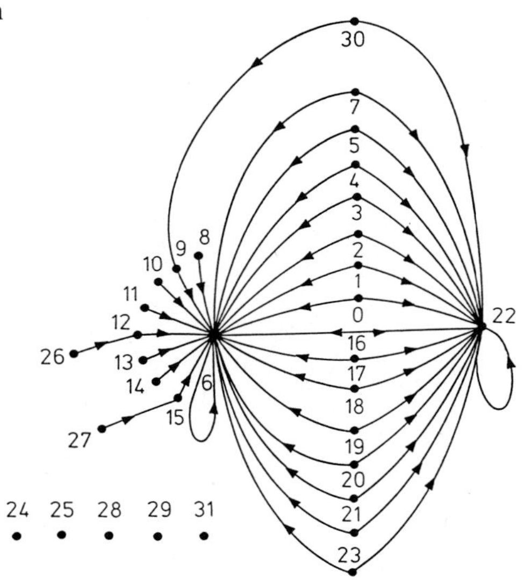

standard Newton $p=2$, $n=6$

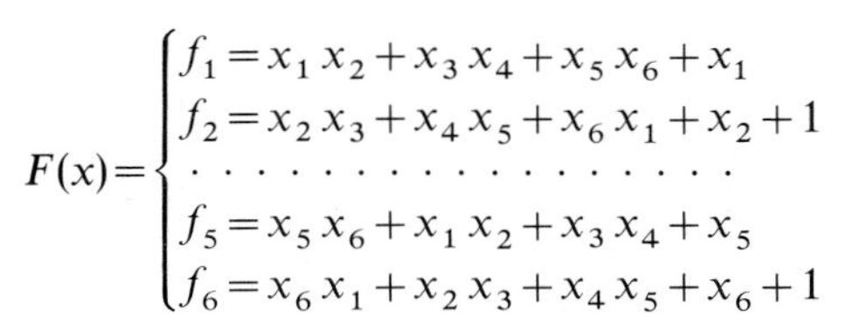

$$F(x)=\begin{cases} f_1=x_1x_2+x_3x_4+x_5x_6+x_1 \\ f_2=x_2x_3+x_4x_5+x_6x_1+x_2+1 \\ \cdots\cdots\cdots\cdots \\ f_5=x_5x_6+x_1x_2+x_3x_4+x_5 \\ f_6=x_6x_1+x_2x_3+x_4x_5+x_6+1 \end{cases}$$

one single root

$21=(0\ 1\ 0\ 1\ 0\ 1)$

in the entire graph

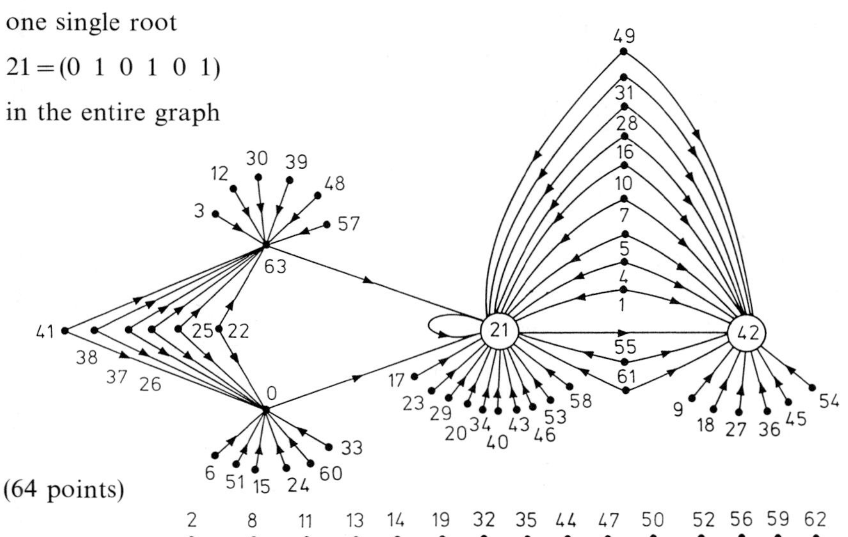

standard Newton $p=2$, $n=8$

$$F(x)=\begin{cases} f_1=x_1x_2+x_3x_4+x_5x_6+x_7x_8+x_1 \\ f_2=x_2x_3+x_4x_5+x_6x_7+x_8x_1+x_2+1 \\ \cdots\cdots\cdots\cdots \\ f_7=x_7x_8+x_1x_2+x_3x_4+x_5x_6+x_7 \\ f_8=x_8x_1+x_2x_3+x_4x_5+x_6x_7+x_8+1 \end{cases}$$

one single root in the partial graph $85=(0\ 1\ 0\ 1\ 0\ 1\ 0\ 1)$

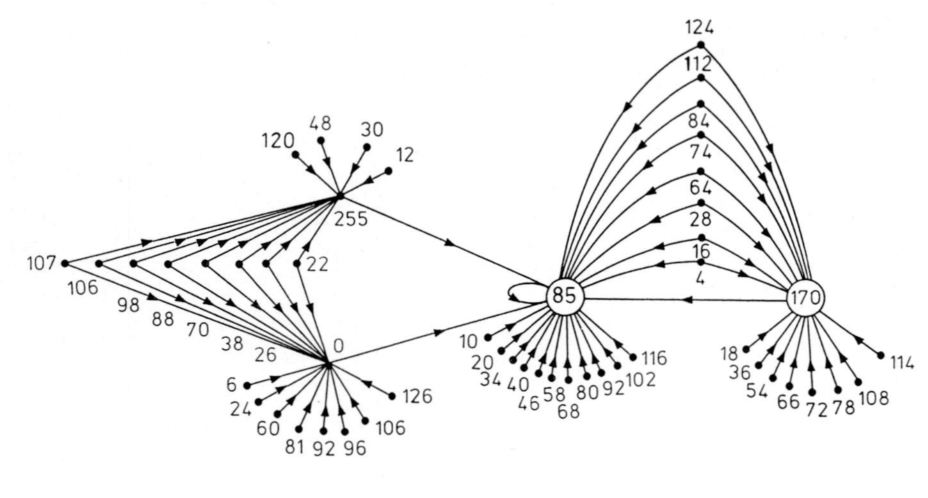

2 8 14 32 42 44 50 52 56 62 76 82 94 110 118 122

(The complete graph has 256 points)

$$F(x)=\begin{cases} f_1 = x_2^2 + x_2 + x_3 \\ f_2 = x_2 x_3 + p - 1 \\ f_3 = x_2 + x_3^2 + p - 2 \end{cases}$$

$$F\colon (Z/p)^n \longrightarrow (Z/p)^n, \qquad p = n = 3$$

standard Newton

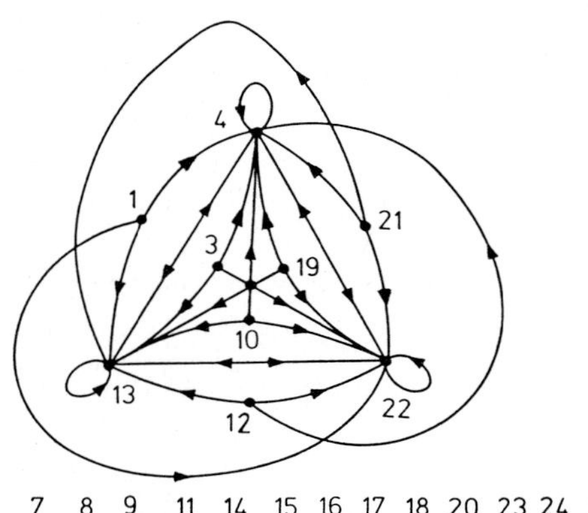

$$F(x)=\begin{cases} f_1=x_1^2+x_1x_2x_3 \\ f_2=x_1x_2+x_2^2+x_2x_3+1 \\ f_3=x_1x_3+x_2x_3+x_3^2+x_3 \end{cases}$$

$F\colon (Z/p)^n \longrightarrow (Z/p)^n, \quad p=n=3$

standard Newton

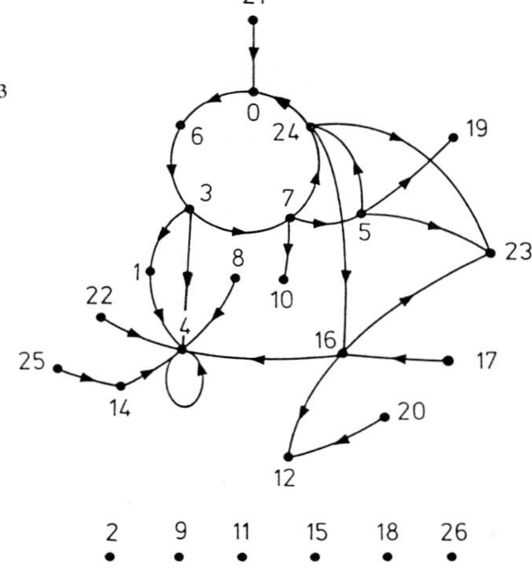

$$F(x)=\begin{cases} f_1=2x_1+x_3 \\ f_2=x_1+x_2+2x_3+x_1x_2 \\ f_3=2x_2+2x_3+x_1x_2x_3 \end{cases}$$

$$A=B=\begin{pmatrix} 2 & 0 & 1 \\ 1 & 1 & 2 \\ 0 & 2 & 2 \end{pmatrix}, \qquad F(x)=Bx+C+U(x)$$

$$C=\begin{pmatrix} 0 \\ 0 \\ 0 \end{pmatrix}, \qquad U(x)=\begin{pmatrix} 0 \\ x_1x_2 \\ x_1x_2x_3 \end{pmatrix}, \qquad F\colon (Z/p)^n \rightarrow (Z/p)^n, \quad p=n=3$$

simplified Newton

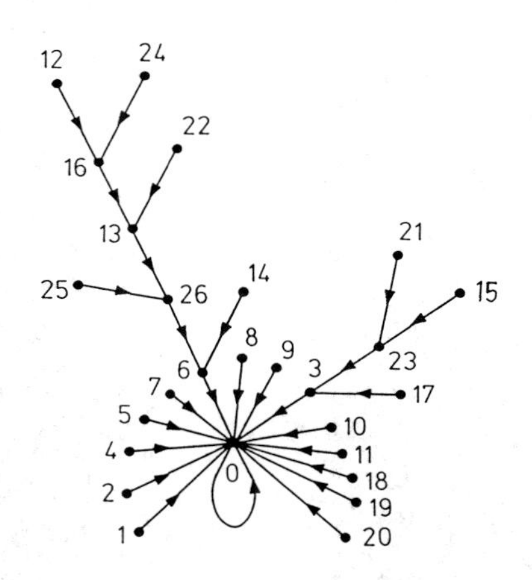

Appendix 4

Continuous Iterations – Discrete Iterations

Consider the following fixed point equation in $\mathbb{R}^2$:

$$\begin{cases} x_1 = \lambda e^{-x_1} + (2-\lambda)\sin x_2 \\ x_2 = (2-\lambda)\cos x_1 + \lambda e^{-x_1} \end{cases} \qquad (x_1, x_2 \text{ in radians})$$

where $\lambda \in [0.9, 2]$.

This equation has an attractive fixed point for all λ given above. The trace of this fixed point is given by the following graph:

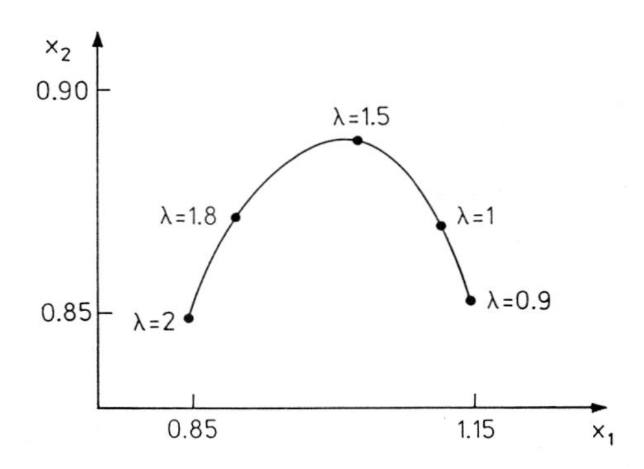

The result of evaluating the functions is now rounded to two decimals. By also evaluating only at values having at most two decimals (the space Z^2) we get a fixed point equation on Z^2 (that is, the map is $Z^2 \to \mathbb{R}^2 \to Z^2$).

For many values of λ (0.9, 0.95, 0.98, 1, 1.1, 1.5, 1.8) we obtain a cyclical behaviour instead of machine convergence as illustrated in the following graphs for $\lambda=1$ and $\lambda=1.8$.

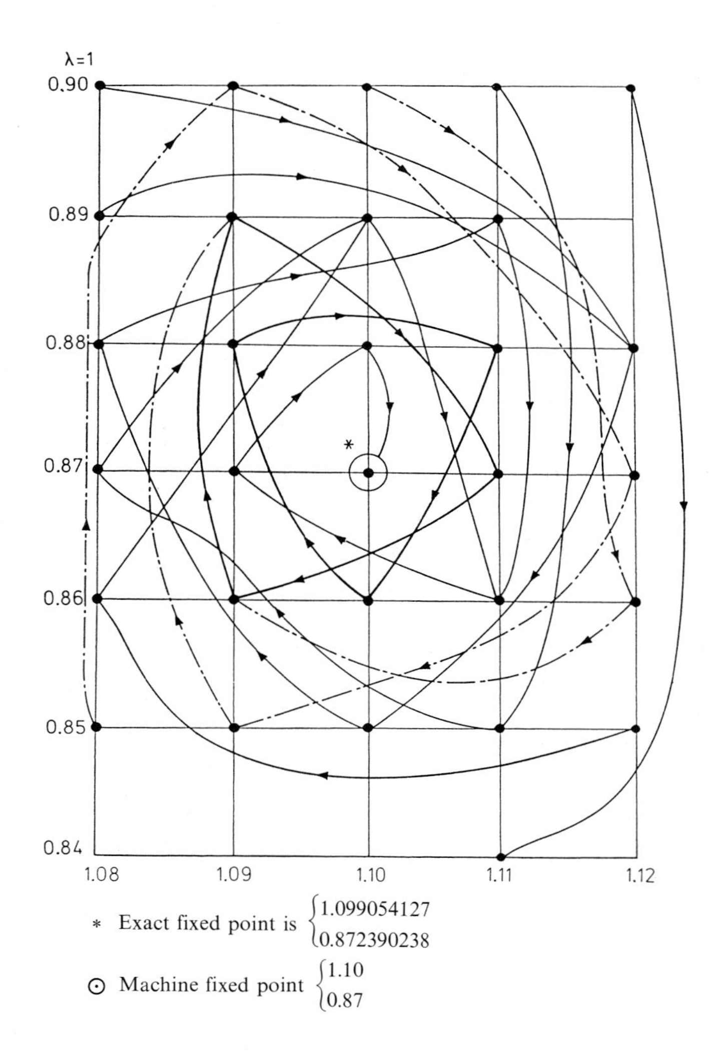

By successive approximations we either reach a fixed point or a cycle, depending on the starting point. The second cycle is isolated.

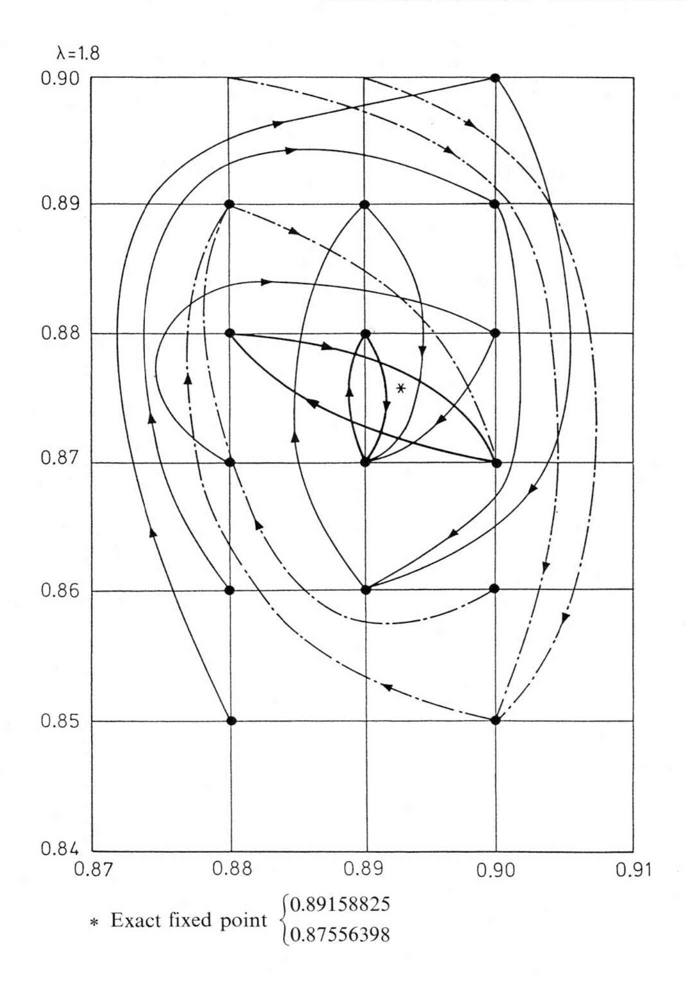

There are no longer any machine fixed points, just cyclical (two cycles) behaviour.

Bibliography

The following bibliography is a collection of references from a number of different areas all having connections with discrete iterations. There is no pretence that the collection is exhaustive!

An attempt has been made to classify these references according to fields of research. This classification might at times seem to be arbitrary. Indeed this is to be expected, given the large number of *interconnections* between the following areas:

A) Continuous iterations and dynamical systems
B) Applied Mathematics: numerical iterations
C) Parallelism in numerical algebra
D) Cellular Automata. Automata networks, Discrete iterations
E) Systolic arrays and VLSI
F) Theoretical Computer Science
G) Biomathematics, neural networks, genetic nets
H) Physics
I) Circuits
J) Boolean matrices and derivatives
K) Psychology, behaviour
L) Miscellanous

A) Continuous Iterations and Dynamical Systems

1. Collet P., Eckman J.P., *Iterated maps on the interval as dynamical systems.* Birkhäuser, Basel (1980)
2. Cosnard, M., Contributions à l'étude du comportement itératif de transformations unidimensionnelles. Thesis, Grenoble (1983)
3. Cosnard, M., On the behaviour of successive approximations. SIAM J. Num. Anal. 16, 2 pp. 300 – 310 (1979)
4. Cosnard, M., Masse, C., Convergence presque partout de la méthode de Newton. CRAS, t 279, pp. 549 – 552 (1983)

5. Coullet, P., Tresser, J., Itérations d'endomorphismes et groupe de renormalisation. C.R.A.S. Paris 287, pp. 577–580 (1978)
6. Feigenbaum, M., The universal metric properties of non linear transformations. J. Stat. Phys. 21 pp. 25–52 (1979)
7. Gumowski, I., Mira, C., *Dynamique chaotique*. CEPADUES Editions, Toulouse (1981)
8. Li, T.Y., Yorke, J.A., Period 3 implies chaos. Amer. Math Monthly 82, 10 pp. 985–992 (1975)
9. Masse, C., L'itération de Newton, convergence et chaos. Thesis, Grenoble (1984)
10. Sarkovskii, A.N., Coexistence of the cycles of a continuous mapping of the line into itself. (Russian) Ukranian Math. J. (16) pp. 61–71 (1964)
11. Targonski, G., *Topics in iteration theory*. Studia Mathematica, Vandenhoeck-Ruprecht, Göttingen (1981)
12. *Transformations ponctuelles et leurs applications*. C. Mira Editor, CNRS Colloquium n° 229 Toulouse Sept. 1973. CNRS Editions (1976)
13. *Théorie de l'itération et ses applications*. Actes du colloque international du CNRS Toulouse (1982)
14. *Traitement numérique des Attracteurs étranges*. M. Cosnard, C. Mira Editors, to appear (Editions du CNRS)

B) Applied Mathematics: Numerical Iterations

15. Baudet, G., The design and analysis of algorithms for asynchronous multiprocessors. PH. D. Thesis Dep. of Computer Science, Carnegie Mellon U. Pittsburg, Penn. (1978)
16. Berman, A., Plemmons, R., *Nonnegative matrices in the social sciences*. Academic Press (1979)
17. Buoni, J.J., Varga, R.S., Theorems of Stein-Rosenberg type, in *Numerical mathematics* ISNM 49, Birkhäuser, Basel pp. 45–75 (1979)
18. Buoni, J.J., Varga, R.S., Theorems of Stein-Rosenberg type II, in *Elliptic problems solvers*. Academic Press pp. 231–240 (1981)
19. Buoni, J.J., Newman, M., Varga, R.S., Theorems of Stein-Rosenberg type III, Linear Algebra Appl. 42 pp. 183–198 (1982)
20. Chatelin, F., *Valeurs propres de matrices*. To appear, Masson, Paris
21. Chazan, D., Miranker, W., Chaotic relaxation. Linear Algebra and its Appl. pp. 199–222 (1969)
22. Ciarlet, P., *Introduction à l'analyse numérique matricielle et à l'optimisation*. Masson, Paris (1982)
23. El Tarazi, M., Contraction et ordre partiel pour l'étude d'algorithmes synchrones et asynchrones en analyse numérique. Thesis, Besançon (1981)
24. Gantmacher, F., *Theory of matrices*. Vol. I and II, Chelsea Pub. Co, (1965)
25. Gastinel N., *Linear numerical analysis*. Academic Press, (1969)
26. Golub, G.H., Meurant, G.A., *Résolution numérique des grands systèmes linéaires*. Eyrolles Paris, (1983)
27. Golub, G.H., Van Loan, C.F., *Matrix computations*. The John Hopkins University Press (1983), 2nd printing (1984)
28. Hageman, L.A., Young, D.M., *Applied iterative methods*. Academic Press (1981)
29. Manteuffel, T.A., An incomplete factorization technique for positive definite linear systems. Math. Comput. 34 pp. 473–497 (1980)
30. Meijerink, J.A., Van der Voorst, H.A., An iterative solution method for linear systems of which the coefficient matrix is a symetric M. matrix. Math. Comput. (31) pp. 148–162 (1977)
31. Miellou, J.C., Algorithmes de relaxation chaotique à retards. RAIRO R1, pp. 148–162 (1975)

32. Ortega, J., Rheinboldt, W.C., *Iterative solution of non-linear equations in several variables.* Academic press (1970)
33. Ostrowski, A., *Solution of equations and systems of equations.* Academic Press, (1960). 2nd edition (1966)
34. Rall, L.B., *Computational solution of non-linear operator equations.* Wiley, New York (1969)
35. Rascle, M., Théorie de Perron Frobenius de certains opérateurs monotones. Thesis, Lyon (1972)
36. Robert, F., Algorithmes tronqués de découpe linéaire. RAIRO, Afcet, pp. 45–64, (1972)
37. Robert, F., Charnay, M., Musy, F., Itérations chaotiques série parallèle pour des équations non linéaires de point fixe. Aplicace Mathematiky (Prague) 20, pp. 1–38 (1975)
38. Robert, F., Matrices nonnégatives et normes vectorielles. Lecture notes, Grenoble (1973)
39. Robert, F., Contraction en norme vectorielle. Linear algebra and its appl. 13, pp. 19–35 (1976)
40. Robert, F., Autour du théorème de Stein-Rosenberg. Num. Math. 27 pp. 133–141 (1976)
41. Robert, F., Machine behaviour of a linear function. (To appear in Computers and Mathematics with Appl.)
42. Saad, Y., Méthodes numériques pour la résolution de problèmes matriciels de grandes dimensions. Thesis, Grenoble (1983)
43. Varga, R.S., Factorization and normalized iterative methods. Boundary problems in differential equations. University of Wisconsin Press, pp 121–142 (1960)
44. Varga, R.S., *Matrix Iterative Analysis.* Prentice Hall (1962)
45. Varga, R.S., Saff, E.B., Mehrmann, V., Incomplete factorizations of matrices and connexions with H-matrices. Siam J. Num. Analysis, 17 (6) pp. 787–793 (1980)
46. Young, D.M., *Iterative solution of large linear systems.* Academic Press (1972)

C) Parallelism in Numerical Algebra: On Overview

See also: [15] [21] [23] [31] [37] above, and Sect. E (Systolic arrays and VLSI).

47. Cosnard, M., Robert, Y., Complexité de la factorisation QR en parallèle. C.R.A.S. Paris, t. 297, pp. 137–140 (1983)
48. Cosnard, M., Muller, J.M., Robert, Y., Parallel QR decomposition of a rectangular matrix. Submited to Num. Math
49. Cosnard, M., Robert, Y., Complexity of the parallel Givens factorization for linear least squares problems. In *Vector and parallel Processors for Scientific Computations,* Ref [62]
50. Evans, D.J., New parallel algorithms in numerical algebra. in *Calcul Vectoriel et parallèle,* Ref. [61]
51. Evans, D.J., Hazopoulos, M., A parallel linear system solver. Int. Jour. Comp. Math. 7, pp. 227–238 (1979)
52. Heller, D., A survey of parallel algorithms in numerical linear algebra. Siam Review 20 pp. 740–777 (1978)
53. Kumar, S.P., Parallel algorithms for solving linear equations on MIMD computers. Ph. D. Thesis, Washington State U. (1982)
54. Lord, R.E., Kowalic, J.S., Kumar, S.P., Solving linear algebric equations on a MIMD computer, J.A.C.M. 30(1) pp. 103–117 (1983)
55. Ortega, J.M., Voigt, R.G., Solution of partial differential equations on vector and parallel computers, ICASE report 85-1, N.A.S.A. (1985)
56. Robert, Y., Tchuente, M., Calcul en parallèle sur des réseaux systoliques. in *Calcul vectoriel et parallèle* Réf. [61]
57. Sameh, A.H., Numerical parallel algorithms: a survey. in *Highspeed Computers and algorithm optimisation.* Academic Press, pp. 207–228 (1977)
58. Sameh, A.H., Kuck, D., On stable parallel linear systems solvers. J.A.C.M. 25, (1) pp. 81–91 (1978)

59. Sameh, A.H., An overwiew of parallel algorithms in numerical algebra. in *Calcul vectoriel et parallèle,* Réf. [61]
60. Tchuente, M., Parallel calculation of a linear mapping on a computer network. Linear Algebra and its Appl, 28, pp. 223 – 247 (1979)
61. *Calcul Vectoriel et parallèle,* A. Bossavit Editeur, Actes du colloque AFCET-GAMNI-INRIA, 17 – 18 mars 1983, Paris. in Bull. EDF, Série C, Vol. 1, (1983)
62. *Vector and Parallel Processors for Scientific Computation,* Academia Nazionale de Lincei, I.B.M. Roma (1985)

D) Cellular Automata - Automata Networks - Discrete Iterations

63. Aladyeff, V., Survey of research in the theory of homogenous structures and their applications. Math. biosciences, 22, pp. 121–154 (1974)
64. Amoroso, S., Cooper, G., The garden of Eden theorem for finite configurations. Proc. Am. Math. Soc. 26 (1) pp. 158 – 164 (1970)
65. Amoroso, S., Lieblein, E., Yamada, M., A unifying framework for the theory of iterative arrays of machines. ACM Symp. Theory Comp. pp. 259 – 269 (1969)
66. Amoroso, S., Guifoyle, R., Some comments on neighbourhood size for tesselation automata. Inf and Control 21 pp. 48 – 55 (1972)
67. Arbib, M., *Brains, machines and mathematics.* McGraw Hill (1964)
68. Arbib, M., A simple self-reproducing universal automaton. Inf and Control 9 pp. 188 – 189 (1966)
69. Arbib, M., *Theories of abstract Automata.* Prentice Hall series (1969)
70. Atlan, H., Fogelman-Soulie, F., Salomon, J., Weisbuch, G., Random boolean networks. Cybernetics and systems 12 pp. 103 – 121 (1981)
71. Banks, E., Universality in cellular automata. IEEE Ann. Symp. Switching theory and automata theory. Santa Monica Calif. pp. 194 – 215 (1970)
72. Burks, A.W., *Essays on cellular automata.* University of Illinois Press (1970)
73. Butler, J.T., A note on cellular automata simulations. Inf and Control, 26 pp. 286 – 295 (1974)
74. Codd, E.F., *Cellular Automata.* Academic Press (1968)
75. Cole, S.N., Real time computation by n-dimensional iterative arrays of finite state machines. IEEE trans. Comput. C 18 n°4 pp. 349 – 365 (1969)
76. Conway, J.H., Berlekampf, E.R., Guy, R.K., *Winning ways for your mathematical plays.* Vol II, Chap. 25 "What is life?" Academic Press (1982). (About J.H. Conway's game "Life", see also Refs. [86] and [101])
77. Cosnard, M., Goles, E., Dynamique d'un automate à mémoire modélisant le fonctionnement d'un neurone. CRAS t 299 1, n°10 pp. 459 – 461 (1984)
78. Cosnard, M., Goles, E., Dynamical properties of an automaton with memory. in *Disordered systems and biological organization* Réf. [191]
79. El Bernoussi, S., Analyse et comparaison d'itérations discrètes; la méthode de Newton dans $(Z/pZ)^n$. Thesis, Grenoble (1982)
80. Fogelman-Soulie, F., Goles, E., Weisbuch, G., Specific roles of the different boolean mappings in random networks. Bull of Math Biology, 44 n°5, pp. 715 – 730 (1982)
81. Fogelman-Soulie, F., Frustration and stability in random boolean networks. To appear in Discrete Applied Math
82. Fogelman-Soulie, F., Weisbuch, G., Goles, E., Transient length in sequential iterations of theshold functions. Discréte Applied Math (6) pp. 95 – 98 (1983)
83. Fogelman-Soulie, F., Stable core in discrete iterations of boolean mappings. In *Dynamical systems and cellular Automata,* Réf. [140]

84. Fogelman-Soulie, F., Parallel and Sequential computation on boolean networks. To appear in Theor. Comp. Science
85. Fogelman-Soulie, F., Contributions à une théorie du calcul sur réseaux. Thésis, Grenoble (1985)
86. Gardner, M., Mathematical Games, Scientific American. 10/11/12 (1970), 1/2/3/4/11 (1971), 1 (1972), 12 (1975)
87. Goles, E., Olivos, J., Periodic behaviour of generalized threshold functions. Comm. Discrete Math. 30, pp. 187–189 (1980)
88. Goles, E., Comportement oscillatoire d'une famille d'automates cellulaires non uniformes. Thesis, Grenoble (1980)
89. Goles, E., Olivos, J., Comportement périodique des fonctions à multiseuil. Inf. and Control 45, (3) pp. 300–313 (1980)
90. Goles, E., Olivos, J., The convergence of symetric threshold automata Inf. and Control 51, (1981)
91. Goles, E., Fixed point behaviour of threshold functions on a finite set. SIAM J. on Disc. Math. Vol 3 (4) (1982)
92. Goles, E., Tchuente, M., Iterative Behaviour of generalized majority functions. Math Soc. Sciences pp. 197–204 (1983)
93. Goles, E., Comportement dynamique de réseaux d'automates Thesis, Grenoble (1985)
94. Goles, E., Tchuente, M., Iterative behaviour of one dimensional threshold automaton. Discrete Math. 81 pp. 319–332 (1984)
95. Goles, E., Dynamics on Positive automata networks. To appear in Theo. Comp. Science
96. Goles, E., Fogelman, F., Pellegrin, D., Decreasing energy functions as a tool for studying threshold automata networks. (to appear in Discrete Mathematics.)
97. Greenberg, J.M., Hassard, B.D., Hastings, S.P., Pattern formation and periodic structures in systems modeled by reaction-diffusion equations. Bull. Am. Math. Soc. 84-6 pp. 1296–1327 (1978)
98. Greenberg, J.M., Greene, C., Hastings, S.P., A combinatorial problem arising in the study of reaction-diffusion equations. SIAM J. Alg.-Disc. Math. Vol 1 pp. 34–42 (1980)
99. Guifoyle, R., Minimum neighbourhood properties of block tesselation automata. Ph. D. dissertation, Stevens Institute of tech. Hoboken, New Jersey (1971)
100. Harao, M., Noguchi, S., On some dynamical properties of finite cellular automata. IEEE Trans. on computers Vol. 27 n° 1 (1978)
101. Hardouin du Parc, J., Paradis terrestre dans l'automate cellulaire de Conway. RAIRO 8 R3 63 (1974)
102. Hayes, B., The cellular automaton offers a model of the world and is a world into itself. Scientific American 250 3 pp. 12–21 (1984)
103. Jiang, Z.Q., Expérimentation des méthodes itératives de Newton et Gauss-Seidel en variables discrètes. Thesis, Grenoble (1982)
104. Kleene, S.C., Representation of events in nerve nets and finite automata. Automata studies, Princeton U. Press (1956)
105. Kosaraju, S.R., On some open problems in the theory of cellular automata. IEEE Trans on computers, vol. 23 n° 6 (1974)
106. Legendre, M., Analyse et simulation de réseaux d'automates. Thesis, Grenoble (1982)
107. Margolus, M., Physics like models of computation. Physica 10 D (1984). Also in *Cellular Automata,* Réf. [139]
108. Moore, E.F., Machine models of self reproduction. Proc. Symp. Appl. Math. Amer. Math. Soc, 14, pp. 17–34 (1962)
109. Myhill, Y., The converse of Moore's garden of Eden theorem. Proc. Am. Math Soc, 14, pp. 685–86 (1963)
110. Pan X.A., Expérimentation d'automates à seuil pour la reconnaissance de caractères. Thesis, Grenoble (1985)
111. Pellegrin, D., Algorithmique discrète et réseaux d'automates. Thesis, Grenoble (1986)
112. Robert, F., Théorèmes de Perron Frobenius et Stein Rosenberg booléens. Linear Algebra and its Appl. 19, pp. 237–250 (1978)

113. Robert, F., Itérations sur des ensembles finis et automates cellulaires contractants. Linear Algebra and its Appl. 29, pp. 393–412 (1980)
114. Robert, F., Dérivée discrète et convergence locale d'une itération booléenne. Linear Algebra and its Appl. 52/53, pp. 547–589 (1983)
115. Robert, F., Une méthode de Newton en variables discrètes. In *Les Mathématiques de l'-Informatique,* Réf. [166]
116. Robert, F., Analogies entre itérations continues et itérations discrètes. In *Théorie de l'-itération et ses applications.* Réf. [13]
117. Robert, F., Basic results for the behaviour of discrete iterations. In *Disordered systems and biological organization,* Réf. [191]
118. Robert, Y., Tchuente, M., On some dynamical properties of monotone networks. In *Disordered systems and biological organization.* Réf. [191]
119. Schultz, H.D., Zellularautomaten zur Bearbeitung der Lokal-Global Problematik. Thesis, Technische Hochschule, Darmstadt (1981)
120. Shingai, R., Periodic behaviour of one dimensional uniform threshold circuits. Trans IEEE Japan, J. 59, A6, (1976)
121. Shingai, R., Maximum period of 2 dimensional uniform neural networks. Inf. Control (41), pp. 324–341 (1979)
122. Snoussi, E.H., Structure et comportement itératif de certains modèles discrets. Thesis, Grenoble (1980)
123. Tchuente, M., Sur la complexité du calcul des permutations sur un réseau en étoile. Colloque RCP Algorithmique, Nice (1980)
124. Tchuente, M., Contribution à l'étude des méthodes de calcul pour des systèmes de type coopératif. Thesis, Grenoble (1982)
125. Tchuente, M., Parallel realisation of permutation over trees. Discrete Math. 39 pp. 211–214 (1982)
126. Toffoli, T., CAM: A high performance cellular automaton machine. Physica 10 D 1&2 pp. 195–204 (1984). Also in *Cellular Automata,* Réf. [139]
127. Toom, A.L., Monotonic binary cellular automata. Probl. Peredachi. Inf. 12 n° 1 (1976)
128. Vichniac, G., Simulating Physics with cellular automata. In *Cellular Automata,* Réf. [139]
129. Von Neumann, J., *Theory of self reproducing automata.* A.W. Burks (Ed.) University of Illinois Press (1966)
130. Waksman, A., A model of replication. J. ACM, Vol 16, n° 1, pp. 178–188 (1969)
131. Wolfram, S., Statistical mechanics of Cellular Automata. Rev. Mod. Phys. 55 n° 3 pp. 601–642 (1983)
132. Wolfram, S., Universality and Complexity in Cellular Automata Physica 10 D, 1 and 2 pp. 1–35 (1984). Also in *Cellular Automata,* Réf. [139]
133. Wolfram, S., Computation theory of Cellular Automata. Comment. Math. Phys. 96 pp. 15–57 (1984)
134. Wolfram, S., Cellular Automata as models of complexity. Nature, Vol. 311, pp. 419–424 (1984)
135. Yaku, T., The constructibility of a configuration in a cellular automaton. J. Comput. Sci. Systems 7, pp. 481–496 (1973)
136. Yamada, H., Amoroso, S., Tesselation automata. Inf. Control, 14, pp. 299–317 (1969)
137. Yamada, H., Amoroso, S., Structural and behavioral equivalence of tesselation automata. Inf. Control, 18, pp. 1–31 (1971)
138. Yamada, H., Amoroso, S., A completeness problem for pattern generation in tesselation structures. J. Comput. System Sci 4, pp. 137–176 (1970)

Some recent Conferences about Cellular Automata and Automata networks:

139. *Cellular Atomata:* D. Farmer, T. Toffoli, S. Wolfram (Eds) North Holland (1984)
140. *Dynamical systems and cellular automata:* J. Demongeot, E. Goles, M. Tchuente (Eds) Academic press (1985)
141. *Computation on cellular arrays. Theory and Applications:* F. Fogelman, Y. Robert, M. Tchuente (Eds), to appear (Non linear science, Manchester U. Press)

E) Systolic Arrays and VLSI

142. Andre, F., Frison, P., Quinton, P., Algorithmes systoliques, de la théorie à la pratique. R.R. Inria n° 214 (1983)
143. Davenport, J., Robert, Y., VLSI and Computer Algebra: the GCD example. In: *Dynamical systems and Cellular Automata,* Réf. [140]
144. Foster, M.J., Kung, H.T., The design of special purpose VLSI chips. IEEE Computer 13, 1, pp. 26–40 (1980)
145. Gentleman, M., Kung, H.T., Matrix triangularization by systolic arrays. Proc. SPIE 298, Real Time Signal Processing IV, San Diego, Calif. (1981)
146. Kung, H.T., Leiserson, C.E., Systolic arrays for VLSI. Sparse matrix proc. 1978 I. Duff and G.W. Stewart (Eds) p. 256–282 SIAM (1979) Also in *Introduction to VLSI systems* Réf. [149]
147. Kung, H.T., Why systolic architectures? Computer magazine 15 (1) pp. 37–46, (1982)
148. Kung, H.T., The structure of parallel algorithms. *Advances in computers* 19, pp. 65–112 (1980)
149. Mead, C.A., Connay, M.A., *Introduction to VLSI Systems*. Addison Wesley (1980)
150. Quinton, P., The systematic design of systolic arrays. in *Dynamical systems and Cellular automata,* Réf. [140]
151. Robert, Y., Quelques algorithmes systoliques pour le calcul scientifique. Thesis, Grenoble (1982)
152. Robert, Y., Block LU decomposition of a band matrix on a systolic array. To appear in Int. J. of Computer Mathematics (1985)
153. Robert, Y., Tchuente, M., A systolic array for the longuest Common Subsequence Problem. *International Workshop on High-Level Computer Architecture*. The Aerospace Corporation, Los Angeles (1983)
154. Robert, Y., Tchuente, M., Calcul en temps linéaire de la plus longue sous suite commune à deux chaines sur une architecture systolique. CRAS t 297, 1 n° 7, pp. 269–271 (1984)
155. Robert, Y., Tchuente, M., Réseaux systoliques pour des problèmes de mots. RAIRO Informatique théorique (1985)
156. Robert, Y., Tchuente, M., Designing efficient systolic algorithms. (To appear in Journal of VLSI and Computer systems.)
157. Robert, Y., Tchuente, M., Reconnaissance de langage en temps réel sur une architecture paralléle spécialisée. CRAS t 300, 1 pp. 363–368 (1985)
158. Tchuente, M., Melkemi, L., Systolic arrays for connectivity and triangularisation problems. In *Dynamical systems and Cellular Automata,* Réf. [140]

F) Theoretical Computer Science

159. Dijkstra, E.W., Self stabilizing systems in spite of distributed control. Comm. ACM, Vol 17, n° 11, pp. 643–644 (1974)
160. Dijkstra, E.W., Guarded commands, non determinacy and formal derivation of programs. Comm. ACM Vol 18, pp. 453–457 (1975)
161. Hinton, G.E., Sejnowski, T.J., Optimal perceptual inference. Proc. IEEE Conf. on Computer Vision and Pattern Recognition, pp. 448–453 (1983)
162. Knuth, D.E., *The Art of Computer Programming*. Vol. 3, Sorting and Searching, Addison Wesley (1973)
163. Nivat, M., Arnold, A., Comportement de processus. In *Les Mathématiques de l'Informatique,* Réf. [166]
164. Sifakis, J., Le contrôle des systèmes asynchrones. Thesis, Grenoble (1979)
165. Tarski, A., A lattice theoretical fixed point theorem and its applications. Pac. J. of Math 5, pp. 285–309 (1955)
166. *"Les Mathématiques de l'Informatique"*. Proceedings AFCET Coloquium, Paris (1982)

G) Biomathematics - Neural Networks - Genetic Nets

See also Sect. D

167. Amari, S., Homogeneous nets of neuron-like elements. Bio. Cybernet. Vol 17 (1975)
168. Changeux, J.P., *L'homme neuronal*. Fayard (1983)
169. Changeux, J.P., Courrege, P., Danchin, A., A theory of the epigenesis of neuronal networks by selective stabilization of synapses. Proc. Nat. Acad. Sci. Vol 70, pp. 974–978 (1973)
170. Cosnard, M., Goles, E., Comportement dynamique d'un automate à mémoire. Analogie avec le fonctionnement d'un neurone. Séminaire de L'Ecole de Biologie théorique, Editions du CNRS (1985)
171. Cull, P., Control of switching nets. Biological Cybernetics 19, pp. 137–145 (1975)
172. Caianello, F., Grimson, W., Synthesis of boolean nets and time behaviour of a general mathematical neuron. Biol. Cybernetics 18, pp. 111–117 (1975)
173. Fikushima, K., Miyake, S., A self organizing neural network with a function of associative memory: feedback-type cognitron. Biological Cybernetics 28, pp. 201–208 (1978)
174. Glass, L., Pasternak, J.S., Prediction of limit cycles in mathematical models of biological oscillations. Bull. Math. Biol., Vol 40, pp. 28–44 (1978)
175. Kauffman, S., Metabolic stability and epigenesis in randomly constructed genetic nets. J. Theor. Biol. 22, pp. 437–467 (1969)
176. Kauffman, S., Behaviour of randomly constructed genetic nets. In *Towards a theoritical biology*, Vol 3, Edinburgh University Press pp. 18–46 (1970)
177. Kauffman, S., The organization of cellular genetic control systems. In *Lectures on Mathematics in the Life Sciences*, J.D. Cowan Ed., The American mathematical Society, Vol. 3, pp. 63–116 (1972)
178. Kauffman, S., Gene regulation networks, a theory for their global structure and behaviours. In *Current Topics in Developmental biology*. Moscona and Monroy Ed., Vol. 6, Academic Press, pp. 145–182 (1971)
179. Kitagana, T., Cell space approaches in biomathematics. Math. Biosciences 19, pp. 27–71 (1974)
180. Little, W.A., The existence of persistant states in the brain. Math. Biosciences 19, pp. 101–120 (1974)
181. MacCulloch, W.S., Pitts, W., A logical calculus of the ideas immanent in nervous activity. Bull. Math. Biophys. 5, pp. 113–115 (1943)
182. Milgram, M., Atlan, H., Probabilistic automata as a model for epigenesis of cellular networks. J. Theor. Biol. 103 pp. 523–547 (1983)
183. Slone, N., Lengths of cycle time in random neural networks. Cornell U. Press, Ithaca (1967)
184. Thomas, R., Boolean formalization of genetic control circuits. J. Theor. Biology, pp 542–563 (1973)
185. Thomas, R., Logical analysis of systems comprising feedback loops. J. Theor. Biology, pp. 631–656 (1978)
186. Thomas, R., *Kinetic Logic*. Lecture notes in biomathematics. Springer-Verlag, Berlin Heidelberg New York, Vol. 29 (1979)
187. Tsetlin, M.L., *Automaton theory and modeling of biological systems*. Academic Press (1973)
188. Vidal, J.J., Les ordinateurs à structure neuromimétique: une perspective d'avenir. Colloque neuro-informatique, Cesta, Paris (May 1983)
189. *Competition and cooperation in neural nets:* A. Amari, M.A. Arbib (Eds) Lecture Notes in Biomathematics n° 45, Springer-Verlag, Berlin Heidelberg New York (1982)
190. *Rythms in biology and other fields of applications:* M. Cosnard, J. Demongeot, A. Lebreton Editors, Lecture Notes in Biomathematics n° 49, Springer-Verlag, Berlin Heidelberg New York (1983)
191. *Disordered systems and biological organization:* Nato Advanced Workshop, les Houches 1985. E. Bienenstock, J. Demongeot, F. Fogelman, C. von der Marlsburg, G. Weisbuch (Eds) To appear Springer-Verlag, Berlin Heidelberg New York Tokyo

H) Physics (Spin Glasses, Frustration, Disordered Systems ...)

192. Angles D'Auriac, J., Méthodes numériques pour l'étude de la dynamique des phénomènes critiques. Application aux verres de spins. Thesis, Grenoble (1981)
193. Barahona, F., Application de l'optimisation combinatoire à certains problèmes de verres de spins. Thesis, Grenoble (1980)
194. Bieche, I., Maynard, R., Rammal, R., Uhry, J.P., On the ground states of the frustration model in spin glass, by a matching method of graph theory. Journal of Physics A, 13, pp. 25–53 (1980)
195. Choi, M.Y., Huberman, B.A., Dynamic behaviour of non linear networks. Physic Review A (1983)
196. Choi, M.Y., Huberman, B.A., Digital dynamics and the simulation of magnetic systems. Phys. Review B (1983)
197. Domb, C., Green, M.S., *Phase transitions and critical phenomena* (I–VI). Ac. Press (1972), Second printing (1976)
198. Hopfield, J., Neural networks and physical systems with emergent collective computational abilities. Proc. Nat. Acad. Sc. USA (79) pp. 2554–2558 (1982)
199. Kirkpatrick, S., Phys. Review B16 pp. 46–30 (1977)
200. Kirkpatrick, S., Models of disordered material. In *Ill-condensed matter* Les Houches (1978), R. Balian & al. Editors, North Holland (1979)
201. Kirkpatrick, S., Gelatt, C.D., Vecchi, M.P., Optimization by simulated annealing. Science pp. 671–680 (1983)
202. Lacolle, B., Sur certaines méthodes de calcul de la Physique Statistique. Thesis, Grenoble (1984)
203. Maynard, R., Rammal, R., Ground state structure of the random frustration model in two dimensions. In *Numerical methods in the study of critical phenomena,* Réf. [208]
204. Nadal, J.P., Derrida, B., Vanimenus, J., Directed lattice animals in two dimensions: numerical and exact results. Journal de Physique (43) pp. 1561–1574 (1982)
205. Peretto, P., Collective properties of neural networks; a statistical physics approach. Biol. Cyp. 50 pp. 51–62 (1984)
206. Shaw, G.L., Roney, K.J., Analytic solution of a neural network theory based on an Ising spin system analogy. Physics Letters Vol. 74 A n° 1, 2 pp. 146–150 (1979)
207. Toulouse, G., Comm. Physics 2 115 (1977)
208. *Numerical methods in the study of critical phenomena.* J. Della Dora, J. Demongeot, B. Lacolle (Eds), Springer-Verlag Berlin Heidelberg New York (1981)
209. *Monte Carlo methods in statistical physics.* K. Binder (Ed) Springer-Verlag Berlin Heidelberg New York (1979)

I) Circuits

210. Benes, V.E., *Mathematical theory of connecting networks and telephone traffic.* Academic Press (1965)
211. Benjauthrit, B., Reed, I.S., Galois switching functions and their applications. IEEE Transact. Comput. Vol. C 25 pp. 79–86 (January 1976)
212. Christensen, B., Galois switching circuits. Int. Symp. on multivalued logic. May 29–31 Morgantown W.V. (1974)
213. Gill, A., *Linear sequential circuits, analysis, synthesis and applications.* McGraw Hill Series in system Sciences (1966)
214. Hennie, F.C., *Iterative Arrays of logical circuits.* Wiley & Sons (1961)

J) Boolean Matrices and Derivatives

215. Bozoyan, S.E., Some properties of boolean differentials and of activities of arguments of boolean functions. Prof. Peredachi Infor. Vol. 14 n° 1 pp. 77–89 (1978)

216. Ghilezan, C., Les dérivées partielles des fonctions pseudo-booléennes généralisées. Discrete Applied Math. 4 pp. 37–45 (1982)
217. Kim, K.H., *Boolean Matrix theory and Applications*. Lecture Notes in Pure and Applied Mathematics n° 70, Marcel Dekker Inc. (1982)
218. Thayse, A., Davio, M., Boolean differential calculus and its applications to switching theory. IEEE Transactions on computers. Vol. C22 n° 4 pp. 409–420 (1973)

K) Psychology, Behaviour

219. Gelfand, A.E., Walker, C.C., A systems theoretic approach to the management of complex organizations. Behavioral sciences 25 pp. 250–260 (1980)
220. Harary, F., Cartright, D., Structural balance, a generalization of Heider Theory. Psychological Reviews, 63 (1956)
221. Polyak, S., Sura, M., On periodical behaviour in societies with symetrical relations. Combinatorica (1982)

L) Miscellanous

222. Busacher, R., Saaty, T., *Finite graphs and networks*. McGraw Hill (1965)
223. Bartee, T.C., Schneider, D.I., Computation with finite fields. Inf. & Control 6, pp. 79–88 (1963)
224. Stone, H.S., *Discrete mathematical structures and their applications*. SRA Computer Science series (1973)
225. Ulam, S., On some mathematical problems concerned with patterns of growth of figures. Proc. Sym. Appl. Math 14, Am. Math. Soc., Providence RI, pp. 215–224 (1962)
226. Ulam, S., *Sets, Numbers and Universes*. MIT Press Cambridge (1974)

Additional bibliography:

Frisch, U., Hasslacher, B., Pomeau, Y., A lattice gas automaton for the Navier-Stokes equation. (Submitted to Physical Review Letters) Preprint LA-UR-85-3503

d'Humières, D., Lallemand, P., Shimomura, T., Lattice gas cellular atomata, a new experimental tool for hydrodynamics. (Submitted to Physical Review Letters) Preprint LA-UR-85-4051

Index

affine 3
aggregation 79, 86, 90, 91
algorithmically (better) 81, 82, 86, 87, 89, 91, 93
analogy 25
approximation (successive) 1, 9
arc 7, 9
asynchronous 78
attractive (cycle) 29, 121, 122, 129
– (fixed point) 29, 103, 107, 109, 129
automaton, automata (networks) 1, 9, 11, 58, 59, 78, 107, 108, 112, 137

basin 2, 3, 4, 11, 57
behaviour (of an iteration) 1, 2
better (algorithmically) 81, 82, 86, 87, 89, 91, 93
binary 10
black box 131
boolean 4
– (matrix, eigenvector, eigenvalue) 43
– (Perron-Frobenius theorem) 43, 51, 58, 114
– (Stein-Rosenberg theorem) 43, 53, 54, 55
– (power) 32, 49, 50, 58
– (spectral radius) 48, 58, 60, 103, 122
– (vector distance) 28, 29, 133
brain 10
bursting 79, 88, 91

cell 9, 108
chain 99, 110
Chazan, D., Miranker, W. 17
circuit 58, 60
clock (pulse) 9
comparison (of operating modes) 69, 79
component 7, 59
condensation (of a vector distance) 40, 60, 62
configuration 1, 9, 108
connected (component) 2
– (graph) 9
connectivity (graph) 7, 58, 60, 108
constant (map) 28, 58
contracting 28, 54, 58, 59, 60, 65, 69, 144
contraction 29, 55, 57, 58, 60, 78, 115
converge, convergence 1, 27, 82, 132
cube (n-) 7
cycle 2, 3, 4, 11, 121
cycling 82

decomposition 17
derivative 117
descendant 2
disconnected (graph) 9
discrete (derivative) 78, 95, 97, 99, 131
– (iterations) 1
– (Newton method) 131
discretization (procedure) 24
distance 27
– (vector) 28, 60, 99, 133

efficiency 132, 153
eigenelements (eigenvalue, eigenvector) of a boolean matrix 43
environment 9

excitization 10
exhaustion 131
expansion (non) 87
expansive (non) 82, 86, 92

fineness (of a discrete metric) 41
finite 1
– (state automaton) 9
fixed point 1, 2, 3, 4, 9, 57, 59, 103, 108, 112, 122
– – (theorem) 58
form (standard – of a boolean matrix) 45
Frobenius (Perron-) theorem 43, 51, 58, 114
function (transition) 9
– composition 110

Gauss-Seidel 11, 25, 54, 65, 69, 70, 76, 115
Gauss-Seidelisations 76
global (results) 95, 101
– (transition function) 9
Goles E. 23
graph 1
– (connectivity) 7, 58, 60, 108
– (iteration) 2, 3, 60
grid 1, 24

hierarchy (of contractions) 72, 74

immediate (neighbourhood) 96, 103, 107, 121, 132, 138
implosion 80, 86, 88, 90, 91
incidence (matrix) 7, 98
inhibition (input) 10
initial (configuration) 9
input (excitization) 10
– (inhibition) 10
iterate (r-th) 9
iteration 1
– (graph) 2, 3, 60
iterative (method) 132

Jacobi 25
Jacobian 95
Jordan (normal form) 45

Kantorovitch 145, 165

length (minimal) 99, 110
– (of a cycle) 2, 121
local (behaviour) 95
– (convergence) 29, 103, 108
– (results) 101
– (transition function) 9

majority (function) 21, 22
– (iteration) 21
massive (neighbourhood) 108, 126, 142
matrix (incidence) 7, 98

method (Newton) 29, 131, 134
– (of exhaustion) 131
metric (discrete) 27
– (space) 29
– (studies) 25
– (tool) 27, 166
minimal (length) 99, 110
Miranker W. 17
mode (of operation) 11
monotone 82, 86, 92
monotonic 153
monotonically 146
monotonicity 80, 87

n-cube 7
neighbour 96
neighbourhood 96, 103, 121, 138
neighbouring (configuration) 108
networks 1, 9, 10
neuron 10
Newton's Method 29, 131, 134
node 1
normal form (Jordan) 45

operating (mode) 57, 69, 79
operation (parallel) 9, 11, 80
– (mode of) 11
optimal (serial process) 70
ordered 82, 86, 87, 92
output 10

parallel (iteration) 59, 65
– (operation) 9, 11, 80
pattern 4, 6
Perron-Frobenius (theorem) 43, 51, 58, 114
process 17
property P 82, 86
pulse (clock) 9

radius (boolean spectral) 48, 58, 60, 103, 122
reducible (matrix) 49
Rosenberg (Stein-) (theorem) 43, 53, 54, 55

Seidel (Gauss-) 11, 25, 54, 65, 69, 70, 76, 115
sequential (more – mode of operation) 17, 57, 72, 79
serial-parallel (mode of operation) 11, 57, 65, 69, 70, 80, 90
shape 17
simple (iteration graph) 6, 28, 57, 60
simplified (Newton method) 134, 138

sparse (matrix) 9, 23
spectral radius (boolean) 48, 58, 60, 103, 122
stable (configuration) 9, 11, 57, 59, 108, 112
standard (form of a boolean matrix) 45
– Newton method 134, 135
starting (configuration) 57
state (automaton) (finite) 9
stationary (convergence) 27, 59, 132
– (iteration) 1
Stein-Rosenberg theorem 43, 53, 54, 55
step 2
successive (approximation) 1, 9
– (Gauss-Seidelisations) 76
synchronous 78

threshold (function) 23
tie (vote) 21
transient 2
transition (function) 9
triangularized 77
truncated (Stein-Rosenberg theorem) 53

vector (distance) 28, 60, 99, 133
vertex, vertices 7

Springer

Springer